SIMON FRASER UNIVERSITY
W.A.C. BENNETT LIBRARY

Work Identity at the End of the Line?

Work Identity at the End of the Line?

Privatisation and Culture Change in the UK Rail Industry

Tim Strangleman

palgrave
macmillan

First published 2004 by
PALGRAVE MACMILLAN
Houndmills, Basingstoke, Hampshire RG21 6XS and
175 Fifth Avenue, New York, N. Y. 10010
Companies and representatives throughout the world

PALGRAVE MACMILLAN is the global academic imprint of the Palgrave Macmillan division of St. Martin's Press, LLC and of Palgrave Macmillan Ltd. Macmillan® is a registered trademark in the United States, United Kingdom and other countries. Palgrave is a registered trademark in the European Union and other countries.

ISBN 1–4039–3980–2

This book is printed on paper suitable for recycling and made from fully managed and sustained forest sources.

A catalogue record for this book is available from the British Library.

Library of Congress Cataloging-in-Publication Data
Strangleman, Tim, 1967–
    Work identity at the end of the line? : privatisation and culture change in the UK rail industry / Tim Strangleman.
        p. cm.
    Includes bibliographical references and index.
    ISBN 1–4039–3980–2 (cloth)
        1. Railroads and state–Great Britain. 2. Privatization–Great Britain. 3. Organizational change. 4. Railroads–Great Britain–Management. 5. Railroads–Great Britain–Employees. 6. Organizational sociology. I. Title.

HE3017.S77 2004
385'.0941–dc22   2004044364

10   9   8   7   6   5   4   3   2   1
13   12   11   10   09   08   07   06   05   04

Printed and bound in Great Britain by
Antony Rowe Ltd, Chippenham and Eastbourne

For Jean and Eric Strangleman – Mum and Dad

Contents

List of Figures   viii

List of Acronyms   x

Acknowledgements   xii

Foreword   xiv

1   From Nationalisation to Privatisation   1

2   Creating Railway Culture, 1830–1947   14

3   On Behalf of the People: Managing a Nationalised Railway
    from Attlee to Thatcher, 1948–79   43

4   A Seat at the Table? Working for the Nationalised Railway,
    1948–79   67

5   Back to the Future? Railway Commercialisation and
    Privatisation, 1979–2001   104

6   A Brave New World: Working for a Privatised Railway,
    1979–2001   134

7   Nostalgia for Nationalisation?   164

Notes   178

References   180

Index   190

List of Figures

2.1 The clerical workforce: staff at Royal Mint Street (Minories) London, Great Eastern Railway Goods Depot, 1898. The author's grandfather, aged 13, is on the front row, far left. (Author's collection)   20

2.2 Corporate pride: the author's great-great-uncle, Robert Strangleman, London North Eastern Railway passenger guard, circa 1923. (Author's collection)   27

2.3 *Railway Responsibility* by John Tenniel. (*Punch*, 1874, Vol. 67, p. 129)   37

2.4 *The Patent Safety Railway Buffer*, Anon. (*Punch*, 1857, Vol. 33, p. 25)   38

2.5 Corporate pride: *The Permanent Way – Relaying* by Stanhope Forbes. Promotional poster for the London Midland and Scottish Railway, 1924. (National Railway Museum/ Science & Society Picture Library)   40

4.1 Footplatemen celebrate nationalisation at Newcastle Central Station, 1st January 1948. (Ken Hoole Collection, Darlington Railway Museum)   68

4.2 *On Early Shift – Greenwood Signal Box, New Barnet.* Promotional poster for BR by Terence Cuneo, 1948. (Cuneo Fine Arts/NRM/Science & Society Picture Library)   75

4.3 *Clear Road Ahead.* Promotional poster for BR by Terence Cuneo, 1949. (Cuneo Fine Arts/NRM/Science & Society Picture Library)   83

5.1 'The new golden age of rail travel is arriving'. Promotional leaflet produced by GNER in 1996. (GNER)   129

5.2 GNER coat of arms, 1996. (GNER)   130

5.3 *Reviving a Legend.* Railtrack promotional poster by Andrew Davidson and David Partner, reproduced in Railtrack Annual Report and Accounts, 1997–8. (Railtrack plc)   132

6.1 *Working in 'Step'.* Railtrack promotional poster by Andrew Davidson, reproduced in Railtrack Annual Report and Accounts, 1996–7. (Railtrack plc)   145

6.2   *Railtrack is Spending £4 million a Day.* Railtrack promotional
       poster by Debbie Cook, reproduced in Railtrack Annual
       Report and Accounts, 1996–7. (Railtrack plc)   150
7.1   *The Momentous Question.* Queen Victoria asks her
       husband Albert if he has any railway shares in the wake
       of the bursting of a speculative investment bubble in 1845.
       (*Punch,* 1845, Vol. 9, p. 45)   168

List of Acronyms

The following acronyms are used in this book.

ASLEF	Associated Society of Locomotive Engineers and Firemen
ASRS	Amalgamated Society of Railway Servants
BP	British Petroleum
BR	British Rail/Railways
BRB	British Railways Board
BTC	British Transport Commission
DFM	Documentary Film Movement
EWS	English, Welsh and Scottish (Railways)
GCC	Gauge Corner Cracking
GDP	Gross Domestic Product
GNER	Great North Eastern Railway
GNR	Great Northern Railway
GWR	Great Western Railway
IEA	Institute of Economic Affairs
LMR	London Midland Region (BR)
LMS	London Midland and Scottish (Railway)
LNER	London North Eastern Railway
LNWR	London North Western Railway
LPTB	London Passenger Transport Board
LT	London Transport
LU(L)	London Underground (Limited)
MIC	Mutual Improvement Class
NER	North Eastern Railway
NUR	National Union of Railwaymen
O for Q	Organising for Quality
RE	Railway Executive
RMT	Rail Maritime and Transport Union
ROSCO	Rolling Stock Leasing Company
SR	Southern Railway
SWT	South West Trains
TCL	Train Crew Leader
TCS	Train Crew Supervisor
TOC	Train Operating Company

TOU	Train Operating Unit
TSSA	Transport Salaried Staffs' Association
WR	Western Region (BR)

Acknowledgements

This book is really the end of a long journey which began in June 1983 when I joined London Transport (Railways) as a 16-year-old junior trainee. I can think of few better training grounds for a sociologist than being thrust into such a complex and fascinating social milieu. One rapidly learnt the 'ins' and 'outs' of a new language of lines, grades and hierarchy, each with their own character and history. For example, Metropolitan signalmen were different from those on the Bakerloo; District-line drivers were a cut above those on the Northern – and so it went on. There was also a divide based on colour of collar. Workers taking the white collar really would start acting differently once promoted and there were often murmurs as to how they had 'changed' as a result. In spite of this complexity and division, this social order worked, and in trying to understand how it worked I became both a railwayman and an untutored sociologist (although I had never heard of the subject). I left the railway in 1988 to go to Ruskin College in Oxford, having been accepted on the strength of a hastily written essay on the pros and cons of privatisation, completed in Canal Junction signal box whilst on early shift. At Ruskin, the world I had left behind started to make academic sense, thanks to tutors such as Peter Donaldson, Harold Pollins, Victor Treadwell and the inspirational Raphael Samuel.

This book is based on research undertaken as part of my PhD at Durham University from 1993 to 1997, as well as other work undertaken since then. In pursuing this study I was very privileged to meet so many kind and generous people. My thanks go to all those members of 'The Club' in Newcastle. I am especially indebted to George Deownly, who so kindly introduced me to a selection of his former colleagues – Morris, Bob, Gilly, Stan, Walter, Tom Richardson, Tiny, Tom Storey, Joe Brown and Harry Friend. I am also grateful to Julie Marr for giving me George's address in the first place. Thanks must also go to all those current railway workers who still work 'on the job', both in LUL and former BR organisations. I hope I have done justice to their views. And I am grateful to Norry Greg for lending me several books that I kept for far too long.

I am grateful too to the former Loadhaul organisation for allowing me access to their company and staff at what was a busy and very difficult period in their short history. Staff at Railtrack, GNER, the Science Museum, the National Railway Museum, the Science and Society Picture Library and Cuneo Fine Arts have kindly given permission for the use of their images in this book. And thanks go to Debbie Cook for the use of the picture on the cover of this volume. I also acknowledge the kind assistance of staff in the Ken Hoole Study Centre at Darlington Railway Museum, and the permission of the Ken Hoole Collection to use one of their photos.

Throughout my period of study I have been helped and supported by a number of colleagues. My thanks go to Pandelli Glavanis, Gill Callaghan, Adam Swain, Dave Wilson, Clive Groome, Colin Divall, George Revill, Andrew Grantham, Mike Savage, Andrew Pendleton, Gerry Hanlon, Jane Lewis and Joan McArthur. Colleagues and friends at Nottingham have been extremely kind and generous with their help and support, so thanks to Meryl Aldridge, Alan Aldridge, Tracey Warren, Alison Pilnick, Brian Rapport, Patrick Wallis and Robert Dingwall. Special thanks must go to Richard Brown for his many kind remarks while reading earlier drafts. I also owe a massive debt to Jane Brown for her love, support and kindness over the years. Ian Roberts was the best and most supportive supervisor one could hope for. Thanks must also go to the ESRC for granting me an award (R00429334335), on which this book is based on. I would like to thank Anneliese Emmans Dean and John 'the toolbox' Smith.

On a more personal note, my thanks go to my Mum and Dad and brother Robin, who have all supported my return to learning in various ways. My love also goes to my partner Claudia, without whose support this work would have been unimaginable. Finally I want to acknowledge the immense debt that I owe to the railway men and women I worked with before embarking on my second chance. I feel very privileged to have become a railwayman in such honourable company, I thank all those I worked with, including Ted, Terry, Phil, Sean and 'Uncle' Ernie at Aldgate, and John Davidson who gave me a push in the right direction at the right time.

TIM STRANGLEMAN

Foreword

There is no doubt that the privatisation of the railways by the Conservative administration of 1992–7 was one of the great government failures of the 20th century. It ranks with poll tax, the winter of discontent, even possibly the Suez crisis as a major policy error brought about by arrogance and the failure by ministers to listen to wise counsel.

The bald facts of rail privatisation are quite shocking: a reasonably well-functioning and economically efficient railway system was broken up into a dysfunctional network that has cost billions of pounds extra in taxpayers' money, increased massively the amount of delays and blighted the lives of many of its workers.

British Rail was by no means a perfect organisation. It was, on occasion, unresponsive to demand and neglectful of the needs of its customers, whether passengers or freight. BR made mistakes in cutting back services and investment on certain crucial routes which later would be shown to be quite wrong. Most crucially, it was subject to the whim of Treasury officials in deciding the level and scope of its investment expenditure. If BR was, at times, too penny-pinching, that was because the government saw, overall, the rail industry as being in long term decline and not as a major contributor to the country's transport infrastructure. But that was a governmental, rather than a BR policy and contrasted sadly with the approach in France and other European neighbours where high speed trains were seen as having an important role in the future.

On the whole, BR was an effective and well run organisation, certainly one of the world's most efficient railways. The number of staff was cut from 600,000 on nationalisation in 1948 to a quarter of that total by the time of its demise half a century later, a remarkable achievement. It had long got rid of its curly sandwiches and it had a forward looking management, made up largely of lifelong railway people who had experience of all aspects of running a railroad. Most sadly, just at the time of privatisation, BR had undergone one of its periodic reorganisations which appeared, this time, to have created the most effective way to run the railway. 'Organisation for quality', as the scheme was known, resulted in considerable power and responsibility being devolved to the various sectors such as InterCity, Network SouthEast, and had met with considerable success. InterCity had become established as a major brand backed by inspirational advertising which had created a renaissance in long distance rail travel. Network SouthEast actually broke even in the late 1980s, a remarkable achievement for a commuter railway mostly used by people in the peak hours.

The impetus for privatisation, therefore, did not come about because of any thorough analysis of the ills of BR or, indeed, of the needs of passengers. It was ideologically driven, spurred on by a Treasury Privatisation Unit which had run out of things to do following the sale of the telecoms and energy industries and therefor looked at the rail industry, with its monopolistic single provider, as an obvious next target. The model, though, as the leading Conservative politician David Willetts now admits, was inherently flawed. As he put it, (*Daily Telegraph* 13 December 2003) 'Rail privatisation was a classic example of taking a model that had worked for one industry and wrongly applying it to different circumstances. We had a model for gas and electricity, where you had a neutral grid, then you had competing providers putting electricity or gas into the grid. The Treasury applied that model to the railways and it was the wrong model.' No matter that a primary school child would have been able to point out that railway is not like an energy grid, not least because trains find it awfully difficult to overtake each other. The model was created for competition between different train companies and that was what drove the way the industry was structured. Yet, ironically, almost as soon as the model began to be applied, the notion of competition proved incompatible with the idea of franchising and was quickly abandoned. The structure, therefore, was created for an aim which was recognised as being unachievable.

Of course this might have been avoided had the politicians and the civil servants listened to those who knew how to run a railway. But they were deliberately shunned. The 30 senior executives in British Rail, mostly people who had worked all their lives in the industry, were called to a ministerial meeting early in the privatisation process and voiced their concerns, unanimously, about the proposed break up and fragmentation of the industry. Instead of recognising this as a legitimate point of view which needed addressing, the ministers, led by the then Transport Secretary John Macgregor, simply categorised the BR managers as Neanderthal and deliberately kept them out of the privatisation process thereafter.

Indeed, BR has subsequently been demonised, with any reference to the past being categorised as harking back to a failed nationalisation.

The facts have never been allowed to get in the way of this analysis. As Tim Strangleman points out in Chapter 6, 'in the railways the implication was that those who looked back to earlier times were seen as nostalgic failures. It was as if established workers were the embodiment of that failure.' Nationalisation is seen as a bad word, and any attempt to analyse the current situation with reference back to British Rail is immediately characterised as harking back to a discredited past. Even now, as the railway is gradually being taken back under state control, ministers are always eager to point out that the industry is not being renationalised.

This ethos of shunning the past quickly percolated down through the whole organisation. Generous redundancy and early retirement packages were offered which made the numbers look good. No value was attached to their experience. Consequently, at privatisation there was a mass exodus throughout the industry, not just of managers, but of people at the shop floor level whose knowledge was costed at zero but worth an almost infinite amount. Those that stayed were made to feel that any reference to the past was inappropriate or mistaken. Strangleman quotes the experience of a former BR man who 'observed how younger but more senior colleagues' eyes would roll every time he tried to make a comment based on previous events or incidents that might have occurred two decades before.

What this book does, for the first time, is to highlight the vital importance of this what is one of those intangible building blocks of civil society which we neglect at our peril, the value of knowledge brought about by experience and a long established culture. The failure of the Tory politicians to listen to the BR managers was repeated time and again throughout the privatisation process. The new railway companies were, for the most part, completely imbued with this 'past is bad, modern is good' culture. As Strangleman puts it, 'What I found most striking when talking to railway workers during the mid-1990s was the way in which railway sense or culture was being actively rejected by the newly privatised companies'. This was across the grades, blue and white collar. ... 'One manager spoke of the way history had almost come to an end when Railtrack was established, likening the position he now found himself in to that in Cambodia after the takeover of the Khmer Rouge.'

The consequences were nothing short of disastrous. Indeed, it is quite possible to argue that several of the railway accidents which happened in the first five years of privatisation were down to the way that this collective but unvalued expertise had been lost from the industry.

To replace it, a whole panoply of expensive safety measures have had to be imposed on the fragmented structure which have pushed up costs of operation to levels at least double those which pertained during the delays of BR. Moreover, much of the damage appears unrepairable. Although the government has begun to recognise that the rail industry has too many interfaces and needs integration, it is difficult to see how this can be achieved without primary legislation and an agenda which starts with a blank sheet of paper, neither of which appears to be on the government's horizon.

To give one example. The escalating costs are a result of the way that decisions about the required inputs, such as track maintenance to get rid of temporary speed restrictions, are made by one set of managers working to certain contractual requirements, while the decision about outputs, such as the timetable, are made by another completely different group. Under an integrated system, not only would these decisions be located in the same organisation, but there would be managers with the required skills to balance the various needs of the engineers, operators and commercial managers and come to rational decisions about what decisions should be taken.

Strangleman documents how the process of privatisation, and the restructuring that led up to it, damaged the skills base in the industry not only because so many people were laid off but also through the systematic undermining of relevant experience through the loss of what he calls 'railway sense' that cannot be replaced by formal training.

These lessons need to be learnt not just in the railways but in other industries, too. This refusal to look at, let alone learn from, the past is a modern phenomenon that is at the root of many basic policy errors. It is, in particular, a New Labour characteristic. In talking to policy advisers and other ministerial acolytes, there is always a sense of dissatisfaction with the task, and an emphasis on reform and change, whatever the particular circumstances. It is almost like Mao's attempt to impose a regime of constant criticism and upheaval, with the old always being less good than the new.

That makes the message of this book of far wider relevance than simply to the railways and those interested in them. These processes are taking place throughout the world of work and while the railways are a particularly poignant and relevant example, there are obvious lessons elsewhere. Change, of course, is necessary and often positive given the rapid advent of new technologies in the 21st century. But so often change appears to be advocated for its own sake. All aspects of the old are lumped together, without any attempt to analyse what should be retained and what should be discarded. The rail industry is a salutary lesson to those who go into this process without sufficient recognition of the value of the past and this book is a warning to them to tread carefully before breaking up a culture simply because they do not understand it.

Christian Wolmar is author of *Broken Rails, how privatisation wrecked Britain's railways*, and *Down the Tube, the battle for London's Underground*, both published by Aurum.

1
From Nationalisation to Privatisation

> 'I could tell you my adventures – beginning from this
> morning,' said Alice a little timidly; 'but it's no use going back
> to yesterday, because I was a different person then.'
> 'Explain all that,' said the Mock Turtle.
> 'No, no! The adventures first,' said the Gryphon in an im-
> patient tone: 'explanations take such a dreadful time.'
> (Lewis Carroll, *Alice's Adventures in Wonderland*, 1998 [1865]: 91)

Introduction

In the early 1960s Clive Groome began work cleaning steam engines
at Nine Elms shed in south London for the then nationalised British
Railways. Dirty but important work, being a cleaner acted as an
important introduction to the world of railway employment. Groome
gradually climbed his way up the promotional ladder, rooted in a
seniority system, as countless generations of railway servants had
before him, to the point where he became a driver, not on his beloved
steam engines, but on more mundane electric commuter trains.
During the 1980s he left the industry because of his increasing disquiet at the levels of monotony, which he believed were dangerous,
and the ever greater management interference in the job of a driver.
After a period of higher education, in the mid-1980s Groome started
his own company called Footplate Days and Ways, which offered
training to those wanting to learn footplate lore. So successful was this
company that he provided training and refresher courses to BR and its
successors for the drivers of nostalgia steam specials on the national
network. In a final twist, the company has developed a profitable side-
line running courses for middle and senior managers at companies

including Ford, on team-building based on the cooperative relationship of footplate legend.

Clive Groome's career acts as a metaphor for what has occurred in the UK railway industry over the last two decades and tells us something wider about the contemporary nature of work and its management. Most directly this story tells about the opprobrium that has been heaped on workers and their managers in the public sector – that they were inefficient, unproductive, overmanned and nostalgically wedded to old-fashioned work practices and traditions. It seems to exemplify the way committed workers have become so estranged from work that they leave jobs they love as working conditions deteriorate and friendships forged over a lifetime are broken up. In another register, Groome's career change could be celebrated as the flowering of the 1980s entrepreneurial spirit that successive Thatcher Governments proudly extolled, freeing up creative talent so readily squashed by the dead hand of corporate government and bureaucratic red tape. Finally, perhaps his story sheds light on a wish in management and Government circles to recapture a supposed lost enchantment in work, a wish to share and replicate a nostalgic vision of what work and workers used to be like when monetary reward was small but occupations offered their own intrinsic satisfaction – in short, the rediscovery of the work ethic.

This book is an attempt to understand the contemporary nature of work and management in Britain, using the railway industry as its main focus. In particular it considers the effects of privatisation, and therefore draws evidence from, and contributes to, far wider debates on the programme of privatisation and the accompanying denigration of the public sector and the service ethic, on the role of management in contemporary Britain, and especially on ideas of corporate culture change. Finally, the book raises important questions about how work is both carried out and also understood by workers, academics and other commentators.

Nationalisation to privatisation

The nationalisation programme of the 1945 Labour Government marked a sea change in policy every bit as dramatic as the privatisation programme four decades later. Before 1939, Governments of both Right and Left had consistently avoided running industry, limiting their involvement to encouraging restructuring and rationalisation, especially in relation to the older traditional industries like steel, shipbuilding, coal and rail. However, within six years of the end of the war a very British mild revolution had occurred, with the State sector employing 2.2 million workers, representing 10 per cent of the UK labour force. The acquisitions by the Attlee administration included:

- British European Airways Corporation in 1946
- the National Coal Board in 1947
- the British Transport Commission and the British Electricity Authority in 1947
- the British Gas Council in 1949
- the Iron and Steel Corporation of Great Britain in 1951 (Saville, 1993).

The next 30 years saw the acquisition of further parts of the private sector, including British Shipbuilders and large parts of the domestic motor industry, so that by 1979 the State industrial sector represented some 10 per cent of GDP, 17 per cent of capital stock and 8 per cent of employment (Rees, 1994: 44).

It would be a mistake, however, to read these events as representing the triumph of socialist ideology on the issue of public control of the commanding heights of the economy. True, there were those within the Labour Party who had always been committed to nationalisation as the first step to a workers' State – an aim of activists in some industries since the late Victorian era. But in reality the Labour Party's long-standing commitment to nationalisation, Clause IV of its 1918 constitution, was framed deliberately vaguely to appeal to both socialists and radical liberals. When the nationalisation programme was enacted, it was as much about economy, efficiency and planning as it was about public ownership per se. And in many ways the actual character of State ownership had been only very sketchily thought through, as is illustrated starkly in a quotation from the autobiography of Manny Shinwell – Minister of Fuel and Power – about the situation he inherited in the coal industry:

> I found that nothing practical and tangible existed. There were some pamphlets, some memoranda produced for private circulation, and nothing else. I had to start with a clean desk.
>
> (Shinwell, 1955: 172)

And Saville notes this was true for most, if not all, of the new undertakings:

> There were few detailed blueprints for the conduct of nationalised industries. The Fabian Society had extolled some ideas, as had the

Labour Research Department, but they were hardly prescribed reading for Ministers, civil servants and businessmen. *The New Statesman and Nation* had much on nationalisation, but little on detail.

(Saville, 1993: 43)

Nationalisation was not, therefore, simply the realisation of a socialist commitment to public ownership made 30 years previously, but rather it also reflected the pragmatic needs of the period and the growing consensus of economic opinion on 'national efficiency' (Tomlinson, 1982; Cunningham, 1993; Saville, 1993; Millward and Singleton, 1995).

The lack of ideological influence on the nationalisation programme is noteworthy and important for understanding why the process took the form it did and, as Tomlinson has suggested, it partly explains its subsequent lack of success. Faced with the problem of how the nationalised industries were to be administered, the Labour Government, in the absence of any other alternative, leant heavily on the concept of the public corporation. This model had proved successful in the prewar period, with its adoption for the London Passenger Transport Board (LPTB) of 1933 (Barker and Robbins, 1974). The aim was to create a unit that was publicly accountable, but not under the direct control of ministers. In addition this solution was favoured because it was *not* representative of interest groups. Saville (1993) has argued that Herbert Morrison's preference was for such corporations to be run by business people and technical experts, rather than by local and national politicians or representatives of the workers employed within the industry.

Tomlinson has criticised this paucity of imagination on so central an issue, arguing that many on the Left mistakenly assumed that the legal form of ownership above all else was crucial to socialised industry. The result was that little if any detailed attention was given to the form and content of the relationship between management and workers within the State-controlled firm (Tomlinson, 1982). Given this lack of a clear conceptualisation about the form public ownership should take, it is not altogether surprising that there was a great deal of continuity between the pre-war private and postwar public undertakings. In many ways this can be seen as the logical extension of the wartime practice of seconding technical, business and labour expertise to industry or Government departments, with sectarian advantage supposedly laid aside in the national interest. This is not to say that nationalisation did not meet with opposition. Many Conservatives and their supporters remained diametrically opposed to it. However, as Saville (1993: 38) puts it:

> the up and coming opposition leaders, including Rab Butler, Harold Macmillan and Anthony Eden, were convinced by electoral defeat that the Tory party had to accept that the war marked a clear divide in British politics. On their return to office all the nationalisations except for steel were accepted as irreversible, even convenient.

By the 1970s this sense in which nationalisation was 'convenient', still less irreversible, was losing its appeal for a growing majority in the Conservative Party. As several commentators have noted, it is important not to impute too much coherence to the Thatcher privatisation programme. It was the Labour Government of 1974–9 that had embarked on the strategy of disposal of State assets, when it sold some of British Petroleum (BP) as part of the loan negotiations with the IMF. Privatisation did not even figure in the Conservative Party's manifesto for 1979. Instead the policy was to emerge gradually over the next decade as *the* centrepiece of Thatcherism. Over the next 18 years, State-run companies were sold off one after another, including Associated British Ports, British Aerospace, British Airways, British Gas, British Petroleum, British Telecom (BT), Britoil, Cable and Wireless, and the electricity, water, coal and finally rail industries.

Wright (1994) argues that this swing against State ownership must be seen as a backlash against the neocorporatism of the post-war settlement. Such a view of the State and its proper role was given a coherence by the economic trouble in which Western industrialised countries found themselves during the late 1960s and early 1970s. Britain, like many developed economies, had problems with growing unemployment, rising inflation and an increasing public sector borrowing requirement. For the neo-liberals now dominating the Conservative Party, the answer was clear: attack the size of the State and 'liberate' industry. Their argument was that State ownership had bred inefficiencies by blurring decision-making between managers, Government ministers, officials, trade unions and workers. A situation had been created in which key sectors of the economy were protected from market forces. The aim was either to return industry to the private sector, or to progressively introduce market pressures and signals into the State sector so as to produce, it was asserted, greater levels of efficiency and productivity.

There are interesting parallels between the postwar Labour and the post-1979 Conservative Governments, in terms of the belief in the efficiency of one sector or another. For the Conservatives, simply moving an undertaking to the private sector was thought to be enough to solve any problems it might have, in much the same way that the earlier Labour administration believed that same efficiency would spring from State ownership. Williams *et al.* (1996: 5) have argued, paraphrasing Orwell's *Animal Farm*, 'Government interference bad, market determination good'. Williams and his colleagues go on to point out that much of the rhetoric of public bad versus private good was simply an assertion based on a perception of poor public services rather than on empirical evidence. Hay (1996) has suggested that the State was portrayed as being in crisis during the 1970s, and in particular during the 'winter of discontent'. This linkage between public-sector crisis and Government failure was instrumental in bringing Thatcher to power.

Such was the faith in the power of the private sector, and the pragmatic wish to see as much of the nationalised sector privately owned as quickly as possible, that the major public utilities – BT, British Gas, water and electricity – simply swapped status from public monopolies to private ones during the 1980s. As Rees (1994: 53–4) puts it:

> The political desire to put a significant privatization programme in place in the lifetime of the first Thatcher government, and to reinforce and extend it in the lifetime of the second, put a tight time constraint on the administrative process of privatization. It was easier and quicker to privatize the monoliths, and even those who were fully aware of the lost opportunities for restructuring … appear to have accepted the argument that it was better to privatize with a sub-optimal structure than to risk losing the opportunity to privatize because of loss of office at an election.

It was only later that greater attention was paid to the question of the creation of competitive markets in the privatised sectors, the most notable example of this being in the rail industry.

Fundamental to a proper analysis of this wholesale move from the public to the private sector is an understanding of the conception of history employed by the Conservatives, and in particular by Mrs Thatcher. Thatcher was at once a traditionalist and a moderniser, Janus-faced like her Party in arguing for revolutionary change in a libertarian sense and yet simultaneously profoundly conservative, particularly in social matters. This tension both united and divided the Tories, its very contradiction providing a platform of difference from which to attack the notion of post-war consensus. And perhaps the best way of understanding this position is in the way history, or more particularly the past, figured in Conservative rhetoric and discourse and how this in turn informed policy. The social historian Raphael Samuel (1998) has argued that the past in the rhetoric of Thatcherism occupied an allegorical rather than temporal space, history used in a very traditional way, as a chapbook by which lessons were learnt and points re-emphasised. Thus the Victorian era was rediscovered and restored as offering a model for a range of social and economic signals for the present – enterprise, hard work, self-improvement, family values and private industry. This restoration was simultaneously marked by the denigration of the postwar settlement with its creation of dependency, slothful bureaucrats, managers and workers, and welfare cheats. As the immediate past was denigrated, so the more distant past took on the aspect of a timeless, enchanted space, where values and traditions are fixed. Thus in seeking to return to the past, Thatcher could present her policy as progressive and modern. Nationalisation became an unfortunate mistake, a lesson from which to learn. Important too was the way management was cast as modern, able to challenge and take on ingrained traditions.

Changing management and culture

This asserted failure of public-sector control was in contrast to the privileging of private-sector management that was now expected to bring about a revolution in the way organisations were run. In his book on managerialism in the public sector, Pollitt (1993: 3) quotes Michael Heseltine speaking in 1980:

> Efficient management is a key to the [*national*] revival ... And the management ethos must run right through our national life – private and public companies, civil service, nationalized industries, local government, the National Health Service.

To neo-liberals, the postwar consensus about the public sector had institutionalised poor management and encouraged restrictive practices which had also infected private industry, sanctioning poor productivity – the so-called 'British disease'. At the same time increasing competition from the economies of the Far East, most notably Japan, seemed to provide a visible and immediate illustration of what had gone wrong with metropolitan capital. The response in the West was in part to question the culture of industry and the workplace. A convenient elective affinity emerged between neo-liberal thought on the economy and a disparate group of management writers who celebrated strong entrepreneurial leadership and the related corporate culture change programmes (see Parker, 2000). The focus of the attack was on bureaucracy, which was viewed as having stifled dynamism and risk-taking within organisations. While bureaucracy restricted both management and workers, it was the former group who were seen as the agents of change who could and should alter the environment of the workplace. One result over the past two decades has been a startling growth in the literature on culture change in organisations, offering various interpretations of the cause and solutions to underperforming cultures.

For the sociologist, the borrowing of social science conceptions of culture in the development of such theories is of particular interest. For instance, there has been a widespread adoption and adaptation of structural functionalist concepts of culture. The attractiveness of such a model for managerialists, and by extension politicians, is the simplistic unitary nature of the problem and the solution: strong healthy organisations have strong healthy cultures; weak organisations have weak cultures (see Lynn-Meek, 1988; Wright, 1994).

Crucially, this elective affinity between neo-liberalism and managerialism coincides in the construction of the past in relation to the present and future. Here, past failure of the established 'old culture' is constructed as the reason for the need for change in practice. Importantly, a new management can develop this critique based on previous management *and* workforce failure, and it is useful to highlight alleged incompetence to reinforce the need for change. Symbolically, then, the past becomes a powerful political plaything within organisations. Opposition to change, be it from the workforce or from within management, can be portrayed as restrictive, conservative and at times nostalgic for a past that has been constructed as having failed. Indeed, that very source of critical commentary from organisational actors is taken as justification for further change, and as such criticism is interpreted as being embedded in the past.

In changing organisational cultures, management is abolishing its own contingency, and to an extent that of the workforce. By this I mean the way that workers acquire resources through their engagement with an organisation over time. Knowledge, custom and practice give individual actors powerful resources that act as a warrant to oppose or support change. Such resources are important ways of mediating the unequal power relations that capital and labour bring to the employment relationship. Rather than having absolute authority, management and supervisors are forced to share power, as work and management relations become embedded and develop. By enacting organisational change, therefore, management are in effect attempting to abolish contingency, changing the architecture of the employment relationship by redesigning the frontier of control between capital and labour. Such culture and organisational changes are an attempt to find the management equivalent of the Holy Grail, namely the ability to start with a clean slate in terms of a workforce and its practices. One way to understand such a move is to see it as an attempt to green a brownfield site. The ability to start from scratch is perceived as the main advantage in a newly established workplace or organisation. Consequently culture change programmes usually involve some parallel structural change which sees numbers of workers – usually older ones – leave the organisation.

Culture change and the contemporary nature of work

Parallel with the debate over culture change has been a discussion about the contemporary nature of work, and in particular the extent to which the changes outlined above, including privatisation, have affected the way work is thought about. Discussions about work-derived meaning and identity are of course not new. During the 1960s there was much debate over the effects of rising affluence on particular sectors of the workforce, and the subsequent marginalisation of 'traditional' workers and their strong occupational communities (Lockwood, 1975). Perhaps what is distinct about the contemporary view of work is the way in which the focus has switched from a collective experience to the level of the individual. In both managerial writing and that by more radical academics there seems to be an account emerging of a loss of meaning and identification with work, or rather there is a sense in which there is a loss of autonomous collective work culture independent of management.

Such is the power that management are supposed to exercise over the workplace that workers now are seen as powerless to oppose culture change initiatives or other new work techniques. This new workforce, then, is seen to be disciplined, or indeed self-disciplined, by a battery of technologies of control, monitoring performance, productivity and attitude. To varying degrees work-based meaning is said to be in decline, being rapidly eroded by a combination of globalisation, industrial restructuring, culture change programmes, de-industrialisation, redundancy and the rise of consumerism.

The writing on this subject could be neatly summarised by the title 'the end of work debate' (see for example Aronowitz and DiFazio, 1994; Casey, 1995; du Gay, 1996; Rifkin, 1996; Aronowitz and Cutler, 1998; Bauman, 1998; Beck, 2000). Although very different in terms of the scale and scope of their research, this group nonetheless share significant similarities in terms of findings. There is a common juxta-position of the importance given to work in the past, both at a societal and an individual level, with the degraded and impoverished nature of current and future employment.

In du Gay's (1996) work, for example, workers are increasingly subject to technologies of power and control; the space for resistance and gaining identity from work is marginal. For Catherine Casey (1995: 2) such is the power of management over corporate culture that:

> The industrial legacy of the centrality of production and work in social and self formation hovers precipitously with the post-industrial condition in which work is declining in social primacy. *Social meaning and solidarity must, eventually, be found elsewhere. [my emphasis]*

For some this situation is one to be celebrated as the chance to rid our-selves of the 'ideology of work'. Gorz (1999: 1), for example, talks about the way society can 'reclaim work' only by ensuring work 'lose[s] its centrality in the minds, thoughts and imaginations of everyone'. Bauman (1998: 17) shares this view, stating that 'work was the main orientation point, in reference to which all other life pursuits could be planned and ordered'. He continues:

> A steady, durable and continuous, logically coherent and tightly-structured working career is however no longer a widely available option. Only in relatively rare cases can a permanent identity be defined, let alone secured, through the job performed.
>
> (Bauman, 1998: 27)

For Beck, too, this transition is equivocal. While the past nature of work is changing or has changed:

> the 'job for life' has disappeared ... paid employment is becoming precarious; the foundations of the social-welfare State are collapsing; normal life-stories are breaking up into fragments.
>
> (Beck, 2000: 2–3)

Like Gorz, Beck believes that such a transition has the possibility for good, in offering a potentially liberating view of fulfilling work in the future.

Richard Sennett (1998) talks of 'the corrosion of character' in the 'new capitalism', and the way the insecurity felt by many has profound implications for moral identity. Here again we have the loss of stability at work having a profound effect on the potential nature of the worker. The implication is that however bad work was in the past, workers could at least develop identity, personality or character on a stable basis.

Why railways?

In trying to make sense of privatisation, culture change programmes and the changing nature of work, this book uses as its focus the railway industry in the UK. This is done for several reasons. Firstly, the railways were the last and perhaps most controversial privatisation enacted by the Conservative Governments of 1979–97. In examining this industry and the way in which it was sold off, it is possible to develop a more sophisticated understanding of privatisation as a whole, and especially the extent to which the Conservatives and their advisors attempted to introduce market forces in this sale to a degree absent in other privatisations. Secondly, the rail industry offers a fascinating account of the management of change over the last two decades. The railways have undergone a series of restructurings and culture change programmes, of which privatisation is only the latest. And thirdly, the railway workforce and the changes it has undergone, especially in the last 20 years, can tell us some important things about the contemporary nature of work, and more particularly about how railway workers continue to identify with their work in spite of the massive changes.

This book seeks to do a series of things. It tells the story of the privatisation of the railway industry, reflecting on the changing political debates over both the nature and scope of marketisation. It looks at how privatisation was enacted, and at the underlying ideas advocating this change. It examines how a mature workforce experienced this change, how it adapted to privatisation and coped with insecurity. But importantly, this contemporary story has to be placed in its historical context, for without a sense of the history of how the industry and its workforce developed, our understanding of its organisational culture can only be incomplete.

Outline of the book

Chapter 2 provides the background to how the railway industry developed in the UK. In particular it looks at the way the structure of the industry was created by the competitive pressures of a multitude of private firms and of Government desire to see competition. It also examines how a new workforce was recruited and managed in such an environment, and how it developed its own distinct traditions and culture. The next two chapters focus on the period of nationalisation from 1948. Chapter 3 examines the way in which Government created the nationalised undertaking, and explores some of the problems and contradictions produced as a result. It is also concerned with the way management dealt with the problems inherited from the private sector, how it sought to rationalise and modernise the system at a time of growing competition with other forms of transport. Chapter 4 then asks how the established workforce experienced nationalisation, and how it responded to the massive changes during this period. The next two chapters examine the period from 1979 to 2002, and focus on the growing concern to make the industry more commercially minded and productive. Chapter 5 concentrates in particular on the way the industry's own past, and its supposed failure, was used as an argument for change, while Chapter 6 examines how the workforce has responded to these pressures. Finally, in the conclusion I attempt to build an analysis around issues of culture change and work identity in the context of the latest developments in the industry.

Method

This book is based on a variety of sources and a mixture of research techniques. Most important are the 55 interviews I carried out with current and retired railway workers, mainly during 1994–6, originally as part of doctoral studies at the University of Durham (see Strangleman, 1998). These were semi-structured interviews with managers, supervisors, trade unionist and shop-floor employees from BR and its successors, and also London Underground. Of this group, one of the managers and a further two 'blue-collar' employees were female. I also undertook nonparticipant observation of a train crew leader in the former Loadhaul freight company (now part of English, Welsh and Scottish Railways), shadowing him for several days, including on a ride on one of the company's locomotives. Since this original research I have carried out further interviews, concentrating on the deteriorating conditions of service experienced by many contemporary workers.

I also make use in this book of the voluminous historical material available on the rail industry, and while there is a danger that some of these sources are 'essays in nostalgia' (see Kellett, 1969), this quality has been useful in and of itself. As will become clear from Chapter 3 onward, I have reflected on this trend in an attempt to understand the wider social significance of railway work, as well as its internal culture. Especially valuable is the vast but historically little used selection of autobiographical writing that has emerged from the industry, most notably during the last 30 years. This has been critical in evaluating the emotional aspects of organisational and industrial change, and the railway historian is particularly fortunate in this respect.

I have also attempted to engage with the writing on organisational culture from within both the social sciences and management literature. I am seeking to combine traditional industrial sociology, which still has much to offer, with material on history, nostalgia, age and generation.

2
Creating Railway Culture, 1830–1947

[I]n London three railwaymen – a guard, an engine-driver, and a signalman – are up before a coroner's jury. A tremendous railway accident has dispatched hundreds of passengers into the next world. The negligence of the railway workers is the cause of the misfortune ... their labour often lasts for 40 or 50 hours without a break. They are ordinary men, not Cyclops. At a certain point their labour-power ran out.

(Karl Marx, *Capital*, Volume 1, 1976 [1867]: 363)

Introduction

In 1844, over twenty years before the accident Marx refers to, J.M.W. Turner completed his famous canvas *Rain, Steam and Speed*. The painting, which now hangs in the National Gallery in London, depicts one of the Great Western Railway's locomotives thundering over a viaduct on the newly constructed route between Bristol and the capital. The enduring appeal of the picture is the way it captures the tension in modernity itself – at once creative and destructive – the brash, technologically advanced railway cutting a swath through a timeless rural landscape, fire spilling forth from its iron belly. The railways brought into being the modern world, they accelerated communication that had hitherto been tied to the pace of the fastest horse, they opened up new markets for goods and services, allowed cities to expand and imposed uniform, 'railway time' on the nation.

The birth of the railways also saw the creation of a new group of workers, an industrial proletariat, who were as much dependent on employment in these increasingly complex organisations as any factory operative. Railway workers occupied a strange occupational space between modernity and tradition, the industrial and the military. Few workers before 1825 were subject to such an intense specialisation in their trades, but railway servants forged new identities and meaning from their work which scarcely any other industries could rival. This chapter explores how it was that railway employment created such strong traditions and loyalties, and seeks to understand the ways in which these distinct customs evolved after 1830. It does this by examining the structure of the industry, the recruitment and training of the workforce and the impact of trade unionism, and questions the extent to which we can talk of a distinct 'railway culture'.

Birth of an industry

The new railway companies were undoubtedly modern organisations. They burst onto a world where most commercial concerns were modest local affairs with few employees, working with relatively small amounts of capital. Around the time Turner completed *Rain, Steam and Speed* in the mid-1840s, the Dowlais iron works in Wales was one of the largest industrial undertakings in the country, with a capital value estimated at £1 million. This was dwarfed by several still relatively new railway companies. For instance, in 1851 the London North Western Railway Company (LNWR) constituted the largest joint stock company of its day, and was capitalised to the value of over 29 million pounds, boasting over eight hundred miles of track linking London with the West Midlands, Liverpool and Manchester (Gourvish, 1972: 108–9).

Such large organisations needed new control structures, and have since been viewed as being in the vanguard of modern industrial corporations. As a result of their sheer scale and geographical spread, railway companies quickly developed centralised bureaucracies, a high degree of management control and a specialised division of labour organised by grades and departments, and the whole organisation was governed by formal law – the company rule book (see Gourvish, 1972; Brown, 1977; Savage, 1998). As Savage (1998: 70) points out:

> Such was the sophistication of the administrative apparatus that in 1919 the GWR had nearly 10,000, or 13 per cent, of its 75,344 staff working in supervisory or salaried grades – a proportion unimaginable in most sectors of the British economy.

The industry grew rapidly during the Victorian era in several cycles of activity created by the railway boom. Each of the 'railway manias' –

occurring in 1837–40, 1845–7 and finally in 1862–5 – was funded by speculative investment obtained from a newly emerging group of eager middle-class shareholders. By the 1870s most of the major towns and cities in the UK were linked by a system of 15,500 route miles, while the finance employed in the industry was equal to one-seventh of the UK's total capital stock. In the latter third of the nineteenth century the network continued to expand, with an eventual total of 20,000 route miles of track.

An important feature of the Victorian railway industry was that the system was planned, designed and built by the private sector, unlike in other continental countries where the State took a far more active role in the construction process (see Stein, 1978; Ferner, 1988). In the UK each new undertaking was self-contained, in that virtually every function was carried out in-house. The companies were vertically and horizontally integrated: they owned and maintained their own track, signalling and infrastructure, and directly employed their own staff throughout the system. Many of the larger concerns also had significant industrial interests. In 1914, for instance, the Swindon works of the GWR employed 12,000 people manufacturing almost everything the organisation required from rivets to locomotives.

Competition was intense, with 467 companies being formed from the first one in 1825 until the Grouping of 1923 (see below). Many went bankrupt or were swallowed up by larger undertakings, but there were still 120 in existence after the 1914–18 war. Although there was some effort at concentration within the industry – by 1874 the industry was dominated by ten giant companies that accounted for 70 per cent of the total mileage and 75 per cent of the gross receipts – Government and public opinion at the time was hostile to the possibility of monopolies being formed by larger companies (Cain, 1988: 103–4).

The State's relationship with the railway industry was complex and at times contradictory. The new sector was born and developed at a time when the *laissez-faire* ideas of the political economists were at their zenith and the free market was widely accepted by public opinion as the only rational way that commerce could flourish successfully (Coats, 1971; Taylor, 1972; Crouzet, 1982: 105). Despite this liberal consensus there was nonetheless sustained Government intervention in the affairs of the industry. Dyos and Aldcroft (1974: 163) note that there were parliamentary select committees investigating some aspect of the industry every year between 1835 and 1840, and then afterwards in 1843, 1844, 1846, 1849, 1853, 1863, 1864, 1872, 1881, 1882, 1891 and 1893. Major legislation regulating railway affairs was enacted in 1838, 1840, 1842, 1844, 1868 and 1889. This apparent contradiction was the result of successive administrations attempting to ensure that the market worked according to classical economic principles rather than being dominated by monopoly power.

State intervention took several forms. Firstly, parliamentary approval was required for the construction and operation of each line, and MPs were reluctant to block the construction of rival routes that duplicated existing services. Secondly, legislation was passed that prevented companies from exploiting a de facto monopoly where there was no effective competition. It did this in a variety of ways, including by fixing rates for individual traffic and preventing companies from refusing custom at a given rate (Milne and Laing, 1956). Finally, and perhaps most importantly, the attempts of the railway companies to merge with one another were consistently blocked in the late nineteenth century. This had a crucial and long-lasting effect on the balance sheets of individual companies, as well as on the economics of the industry as a whole. If the high cost of construction and operation ensured that barriers to entry in the industry were high, the cost of exit was also substantial. The railways, as we have seen, were immensely capital intensive with very high rates of fixed capital and hence fixed costs. Much of this capital was sunk in the sense that it was spent on the viaducts, bridges, embankments and earth works on which the rail was laid, and these assets had little or no residual value. Although early companies made high returns, which fuelled further speculation in the industry, by the 1860s and 1870s these became much smaller, with many undertakings often failing to pay dividends. This meant that the industry as a whole was unable to invest sufficient capital in new technology in the long term because it failed to realise viable economies of scale. It also had important consequences for the railway workforce, since it was in this area that railway companies could – and did – economise. With so much of the industry's costs being fixed, the wage bill was by far the greatest variable cost. But even here flexibility was restricted by State intervention in the hours of railway servants and in the introduction of compulsory safety devices (see Irving, 1976, 1978).

Recognition of the problems within the fragmented system was behind the State-backed concentration of the railways after the 1914–18 war. In 1922 the Government passed legislation which effectively forced the industry to rationalise by grouping some 120 undertakings into four regionally-based companies: Great Western Railway (GWR), Southern Railway (SR), London Midland and Scottish

Railway (LMS) and the London North Eastern Railway (LNER). These amalgamations (which came into effect on 1 January 1923) were the post-war Government's response to several problems. The war had brought about structural changes and efficiencies that politicians were reluctant to lose in peacetime (Crompton, 1995: 116). There had also been Government-imposed wage and hour regulation on parts of the industry's workforce during the war that the smaller companies would have found difficult to adjust to with a return to the *status quo ante*. In addition there was increasing competition from municipal tramways and road transport which threatened profitability, further weakening the ability of the industry to invest (Pollins, 1971; Bagwell, 1974; Irving, 1976).

The preferred Government option for concentration of the industry was grouping rather than outright nationalisation, although this was considered and had been permissible under Gladstone's Railway Act of 1844 (Bagwell, 1974; Foreman-Peck and Millward, 1994; Crompton, 1995). The grouping strategy adopted was not surprising given the overall lack of State involvement in running industry generally.

While the Government had recognised the need to restructure the industry in 1923, it still did not attempt to tackle some of the historic obligations placed on the railways with regard to the carriage of certain traffics or the strict regulation of price (Milne and Laing, 1956). The industry was neither a true monopoly, nor was it competing on a level playing field with its new and growing rivals. The problem for the railway was that it had to cope with this new source of competition and restructure itself to achieve the expected economies of scale against the background of the deep recession of the 1920s and 1930s. The LNER was perhaps the most badly affected of all the new companies, because of its historic reliance on serving the areas of heavy industry in the north-east of England. Recessions in the coal, shipbuilding, iron and steel industries took a serious toll on traffic levels (North, 1975; McCord, 1979). The rail industry as a whole saw a decline in its traffic in both freight and passenger journeys. In 1923, total freight carried amounted to 343.3 million tons,[1] compared to the 1932 figure of 249.6 million tons. Although figures improved during the 1930s, they remained under the 300 million ton mark, and indeed were falling before World War II. The depression saw total passenger journeys down from over 2,186 million in 1920 to a low of 1,231 million in 1935 (Pollins, 1971: 156–7).

In 1939 the railways found themselves again under Government control, with their shareholders paid a fixed sum in return for the use of their assets. The industry suffered greatly during World War II, unlike in the previous war. This was due both to damage from bombing and the lack of maintenance. Although the companies reverted to their pre-war organisation in 1945, it was clear that reorganisation was again inevitable because of the massive cost of repairs and the new investment needed. In 1948 the whole industry was nationalised, becoming British Railways (BR), itself part of a wider British Transport Commission (BTC) responsible for all inland transport (Bonavia, 1971; Pollins, 1971; Gourvish, 1986).

Creating a workforce

The demand for workers in such a large industry was immense. Before 1824, railway employment as an occupational category simply did not exist; by 1847 there were 47,000 staff, increasing to 112,000 in 1860, and 275,000 by 1873 (McKenna, 1976: 26). At nationalisation in 1948 there were 641,000 people employed by the newly created British Railways (Gourvish, 1986: 99). During the peak of construction in 1846–7 it is estimated that over 200,000 men were at work (Brooke, 1983; Coleman, 1986). These 'navvies' gained a fearsome reputation. Coleman (1986: 105) quotes Robert Rawlinson, an engineer involved in the construction of the London and Birmingham Railway during the 1830s on the character of the navvy:

> Nothing would keep these men from the feasts, and joining in revelry and drunkenness, as long as they had a farthing to spare.

In contrast, once the line was completed, railway work was often highly sought after as a source of respectable and permanent employment. As Howell (1999: 3) says:

> Their security was often assured. The slogan that 'a job on the railways is a job for life' remained true, despite the inter-war depression and the impact of road competition, until the advent of mass closures and rapid technological change in the 1960s.

These new workers were an important part of the Victorian uniformed working classes, quickly becoming a byword for stability in urban and rural settings alike. The new railway companies demanded and created a highly differentiated workforce to operate their systems. As one of the keenest observers of the new capitalist system, Marx believed that

Figure 2.1 The clerical workforce: staff at Royal Mint Street (Minories) London, Great Eastern Railway Goods Depot, 1898. The author's grandfather, aged 13, is on the front row, far left. (Author's collection)

the greater division of labour cruelly distorted an individual's humanity. He wrote in the first volume of *Capital*:

> It converts the labourer into a crippled monstrosity, by forcing his detail dexterity at the expense of a world of productive capabilities ... Intelligence in production expands in one direction, because it vanishes in many others.
>
> (Marx, 1954 [1867]: 340–1)

Other classical theorists reflected on this tension that 'modern' work was both destructive and creative of identity. For Weber, the greater specialisation in profession and occupation created new identities, but these were necessarily depersonalised and rationalised (Weber, 1964; Eldridge, 1970). Employment in the railway industry is a perfect illustration of this contradiction. Often recruited from rural agricultural areas where the pace of work was governed by the hours of daylight and the seasons, new workers rapidly found themselves part of a highly developed and complex division of labour, based on grades and departments. But McKenna (1976, 1980) and others have suggested that, far from destroying character, railway employment forged both new skills and identities amongst its workforce.

By 1870 there were nearly one hundred separate grades of railway work (Kingsford, 1970: xiii). Pre-eminent among these early groups were the footplate men, the driver and fireman, whose relative skill levels ensured they received higher levels of pay. Indeed Eric Hobsbawm (1986) included some of the drivers in his category of the super-aristocracy of labour. Certainly the Victorian engine driver embodies Hobsbawm's six characteristics of the labour aristocrat:

1. level and regularity of wage
2. prospects for social security
3. conditions of work
4. relationships with those above and below
5. general conditions of living
6. opportunities for advancement.

For other railway employees, though, wages and conditions were seldom as generous. Kingsford (1970: 101–2), using figures from the London, Brighton and South Coast Railway, noted the difference between the rate of pay for engine drivers of 39 shillings per week and that for a station porter who might only realise a wage of 16 shillings and 4 pence. The majority of workers earned much less than the driving grades, and indeed most enjoyed less than twenty shillings a week.

It is difficult to generalise about the way the companies recruited, trained and employed their staff, especially in the early years of the industry. In essence the nascent industry had no direct models to follow and its labour policies were either borrowed from other large organisations, most notably the military, or simply evolved by trial and error. The recruitment of army and naval personnel ensured that railway work occupied a borderland between the civilian and the military, and Revill (1989) demonstrates that there were frequent references to the 'railway army' in nineteenth-century texts on the industry. The attraction in using personnel from such backgrounds was, writes Bonavia (1971: 13):

> the lack of experience, outside the Armed Forces, of the arts of controlling and disciplining large bodies of staff.

Bonavia argues that this heritage left a permanent mark on the industry in the continued use of military language down to the 1960s and 1970s:

> Their managers were 'officers', their luncheon rooms 'messes'. The rule book was the equivalent of King's Regulations. Discipline ... was strict, with punishments and commendations recorded on every employee's history sheet.
>
> (Bonavia, 1985: 152)

The popular view among railway historians is that a bureaucratic recruitment and promotion system based on seniority rather than patronage was quickly established. Mike Savage (1998) in his discussion of the notion of the 'career' in railway work argues that this formal system was absent in many lines until much later than is often supposed. He is critical of the way authors such as Kingsford and McKenna give the impression that career trajectories were formalised from very early on, with workers joining a particular department and climbing the organisation vertically during their careers. His own work on career progression in the Great Western Railway suggests that:

> In fact, the types of career routes which workers had open to them, and the 'hoops' through which they had to pass to move into

better-paid employment were subject to major change, especially in the period between 1860 and 1900 ... Career routes were not systematized, and a range of job movements was possible, there often being no clear ladders connecting jobs together.

(Savage, 1998: 72)

Evidence from the North Eastern Railway (NER) supports this alternative reading. For example, the company's employees could enjoy a range of posts as well as extremely rapid promotion during the 1860s and 1870s. The NER staff magazine offers the following notices of retirement in 1914 and 1913 respectively from workers who joined the industry in 1860. Robert Nicholson, who was eventually to become a locomotive inspector with the NER, had started as a platelayer in 1860, became a cleaner in 1866, a fireman in 1867 and in turn a driver in 1870 (*NER Magazine*, 1914, Vol. 4, Issue 44, August: 187). J.G. Brown of Tweedmouth shed was made fireman in 1869, driver in 1872 and later still went on to be a foreman (*NER Magazine*, 1913, Vol. 3, Issue 25, January: 20). The rapid promotion seen in the cases of Nicholson and Brown reflects careers that coincided with the expansion of the system in the latter half of the Victorian era. This kind of advancement tended to slow as the industry matured, meaning that workers had to wait longer and longer to gain the higher-paying senior posts. Irving (1976) suggests that frustration at such delay helps explain the labour unrest in the industry at the turn of the century. Contrast the rapid promotion of Brown and Nicholson with the career histories outlined in Hollowell's (1975: 233–4) research, in which he showed that the average time taken to move from cleaner to driver among workers joining the locomotive department between 1916 and 1920 was 25 years. Even those recruited in the period 1946–50 had to wait 16 years to achieve the status of junior driver.

The labour policies adopted by the different companies converged towards the end of the nineteenth century. Promotion among blue-collar workers became increasingly dependent on seniority, vacancies being filled by the senior applicant in the grade or company, underpinned by formal rules based on length of service. Such a system was buttressed by the department structure, which by the turn of the century offered little possibility of cross-transfer once a worker had stepped onto a particular occupational ladder. Some commentators have argued that the use of seniority as a promotional criterion was itself a control strategy. In his discussion of North American railroad practice, Gratton (1990) suggests that seniority, coupled with other forms of security, was an effective way of retaining skilled workers by cutting labour turnover, an acute problem in the USA at the time. As he explains:

> A more coherent rationale for the Pennsylvania's labor policies lies in a deeply embedded system of compensation by seniority. Unlike distant pensions, rewards for tenure offered tangible lures to workers of every age; they encouraged career-minded applicants to seek railroad work and discouraged turnover among current employees. As 'standard corporate policy' on major railroads by the late nineteenth century, seniority-based compensation was not an imposition of unions, but a management tool for securing a stable and experienced labor force ... The recourse to seniority rights revealed central management's desire to decrease turnover, to cultivate an expectation of fair treatment from loyal workers, and to restrain the 'arbitrary power' of local officers.
>
> (Gratton, 1990: 634)

Savage (1998) likewise argues that the construction of the notion of 'career' by railway companies was an alternative to direct control mechanisms, the subjective identification with one's own career effectively becoming a self-disciplining mechanism (see also Grey, 1994). Revill (1991) has written of the way workers were effectively 'trained for life', in that they often developed a high degree of company- (and industry-) specific knowledge that had little practical use in the external labour market. The great advantage railway employment had over the alternatives was the level of security combined with the chance of advancement. While wages may have been low for many, few other contemporary organisations could match the railways for prospects of promotion. Companies were often paternalistic towards their staff in a variety of ways, many building company housing, providing company doctors, and occasionally pension schemes. But the most direct form of welfare was the ability of these large horizontally and vertically integrated corporations to accommodate those who had suffered accident or illness that prevented them from working in their original posts.

Paternalism, however, had a darker side. If job security and good prospects were one side of the coin, the other face was an autocratic management style and a dependent workforce, whose members could be dismissed for relatively minor offences. Although the railway companies needed highly trained, occupationally specific employees at a minimum cost in terms of wages and turnover, they could use the threat of redundancy to discipline workers. As Frank McKenna (1976: 32) wrote:

> On joining the railway and signing his Rule Book, the Victorian railwayman established a covenant with his employer. Within this covenant was security. Outside it – in the conditions of Victorian life – was chaos ... The company promised him a job for life, if he did not offend against its rules, and a position which was likely to improve, through the system of promotion, in stark contrast to other occupations where many men were old at forty. In Hobbesian terms, railwaymen surrendered their freedom in return for security. Thomas Hobbes' contention that self-preservation is the dominant impulse in men is quite clearly demonstrated in the way that thousands of Victorian working men were ready and eager to accept these terms, and to commit themselves to a company for life.

For all grades, the very security at work was underpinned by a fear of what lay outside, a point reinforced by Stedman Jones (1992: 53) who wrote of the Victorian London labour market:

> Very few workers could expect a working life of stable employment in the nineteenth century, and occupations which appeared relatively immune to the hazards of seasonality, cyclical depression, or technological development – brewery or railway employment for instance – were eagerly sought after despite indifferent wage rates.

At times railway management used such dependency in quite sophisticated ways in order to control the workforce, while at other times threats were as crude as in any industry. In his collection of writings on railway labour, David Howell (1999) charts the growing disquiet felt by workers at the threat posed to such stability in the late nineteenth century by the arbitrary exercise of power by management. This tightening of control was the result of intense competition for traffic between companies, a situation made worse as the industry's expansion slowed after 1880. One company that surfaces frequently in Howell's narrative is the Midland Railway, which suffered from growing pressure from its rivals, forcing it to take drastic measures simply to survive. Such retrenchments meant deterioration in the conditions for staff, and resulted in the breaching of 'conventional expectations about the security of railway employment' (Howell, 1999: 91). What made matters worse on the Midland were the actions of

Cecil Paget, the company's young general superintendent, who almost single-handedly waged a campaign against unionisation. Howell tells how after one driver had previously criticised the Midland during a parliamentary enquiry into the conditions of railway servants, Paget leapt onto the moving footplate of the driver's train and accused the man, still attempting to drive his locomotive, of being 'a disgrace to the company's service and the sooner you are out of it the better' (Howell, 1999: 94). Such was the menace behind this threat that the driver, with 19 years' service on the Midland, left the railway industry altogether rather than wait to be sacked. In another case involving the same company, a driver was fired for arguing with a foreman. When his appeal failed, the victim was privately told by a manager sympathetic to his cause that his only hope of exercising his craft further was to apply for a position in Chile. Such was the power of these Victorian industrial Leviathans that they could banish those who offended against them, ensuring transgressors never worked in Britain again.

While there was a level of stability for a core element of the workforce, the railway companies also made limited use of part-time and casual workers to cope with day-to-day or seasonal swings in traffic. There is a passage in the minutes of an 1891 parliamentary select committee sitting on railway servants' hours of labour in which evidence was offered by the NER as to the difficulty of fixing regular hours for its workers.[2] Tennant, the NER General Manager, stated that the amount of iron ore traffic through Sunderland dock could vary between nil and 6,000 tons per week, and that coal loads shipped through Tyne dock in Newcastle could vary between a daily low of 6,000 tons and a high of 24,000 tons. To restrict the company's ability to use its workforce as it wished, Tennant argued, was to endanger company profit margins. The opposition was ultimately unsuccessful and Parliament regulated the hours of railway employees from the last decade of the nineteenth century.

Not surprisingly, it was the unskilled grades that were to suffer the most from this form of instability, with labour being recruited and released as traffic levels dictated. But this type of employment would often act as an entry point to a more secure position once a vacancy arose in the 'core' workforce. This was a strategy employed by railway management well into the twentieth century, with several interviewees who began their careers during the 1930s telling how they were hired in this way (see Strangleman, 1998). Redundancies in the full-time workforce were avoided by using the grades structure to step up and step back labour to avoid employing too many permanent workers.

Thus, if there was a need for an extra driver for one shift, a passed driver would step up to cover the position,[3] and he in turn would be covered by a fireman, who would be covered by a cleaner, and so on. Adoption of such a system avoided large-scale redundancy and recruitment drives among the railway companies until the interwar depression (see Bagwell, 1963).

Figure 2.2   Corporate pride: the author's great-great uncle, Robert Strangleman, London North Eastern Railway passenger guard, circa 1923. (Author's collection)

Whatever the faults of the private companies, there can be little doubt that they created strong identification with work, either directly or indirectly. Howell (1999: 3) argues:

'The Company' became a central feature of many railway workers' lives, and loyalty was assiduously promoted by managers and supervisors. Such identities often proved attractive. For years after the amalgamations of the early 1920s, railway workers retained an affinity with their former company.

But company identity was simply one source of meaning for railway workers, and it is to some of the other sources that we now turn our focus.

Training and socialisation

Once in a company's employment, individuals would find themselves immersed in a Byzantine caste system of competing identities based on region, company, grade and even locomotive shed or station. The railway nation therefore came to be 'inhabited and worked by sections of staff as easily identified as settled tribes' (McKenna, 1976: 66). Arguably McKenna's most important contribution to our understanding of railway labour is his description of the 'railway bailiwick', a defensible space within which workers formed identities, customs and habits with little direct supervision from the company's managers and supervisors. This profound sense of autonomy over work is centrally important to understanding how occupational identities in the industry were formed and reproduced over time. Nowhere is this independence more strikingly illustrated than in the context of training and socialisation. Railway companies provided little in the way of formal training; instead they expected new entrants to learn their trade as they progressed in their careers – indeed they relied on them to do so. As Michael Reynolds stressed in *The Engine-Driver's Friend*, first published in 1881:

The science of locomotive-driving is based upon experience and observation. Things appertaining to the foot-plate find their place in men's experience, and but little in books.

(Reynolds, 1968 [1881]: 16)

Such reliance was an important feature of the industry insofar as it created a highly autonomous workforce with strong traditions and customs associated with self-help and mutual improvement. It is difficult to separate training and socialisation, as Penn (1986) also found in the context of skilled engineering trades. Young railway workers developed technical knowledge alongside work and grade identities. Arthur Turner, who eventually became a driver, describes his early experience of such shed work for the LMS in Bath:

> When engines had been stopped for boiler washing-out, general repairs or just because they hadn't been used for a while, the firebox would have to be cleaned from the inside. This meant the bar boy had to get in through the fire-hole feet first and with head downwards, then with special tools clean the firebox of all rubbish that could not readily be done from outside. All fire-bars would have to be examined and changed, if necessary. This task was one of the least pleasant as it was a very dirty, hot and unthankful job ... All this was hard work for a young lad, but it was taken as part of the job, and a good basic training for future days.
>
> (Turner, 1996: 9–10)

Frank McKenna, who started his career as an engine cleaner in the same way as Turner had, described the complex social order even within the most junior members of shed staff, with each step in the hierarchy marking the acquisition of new skills and status:

> Sweating with fear as well as activity I climbed up into the belly of the engine across brake hangers, stretchers and slide bars. I sat across a brake stretcher, the filth oozing through my overalls and trousers, trapped in space. For more than an hour I perched there in the darkness broken only by the spluttering of the flare lamp, almost overpowered by the sense of isolation and danger. The dankness of the cotton waste, the sickly smell of the cleaning oil and the menacing black underbelly of the boiler petrified me ... I had now discovered the daily ritual of the junior cleaner on the shift.
>
> (McKenna, 1980: 118)

Although never given the status of a formal apprenticeship, the career ladder in this and other grades prepared younger workers in a variety of ways for later, more responsible jobs. Technical skill, as outlined by Turner and McKenna, was overlaid by social skills associated

with railway work in general, and the identity of the grade in particular. From the post of cleaner, young men in the locomotive department would be drawn on to the footplate. Walter Mulligan describes
his attitude to the drivers above him when he started firing in the
1940s:

> When I first started though you had the 'old school' type of driver.
> 'Two paces behind me son.' And what they said was law ... They
> wouldn't even lift the lamp off the train ... some of them. They
> were there to drive the train and that was all they did.
> (Walter Mulligan, interviewed 1995)[4]

Former footplate workers would often talk about the way they had to
'prove' themselves before they would be fully accepted by established
drivers during the 1930s and 1940s. Such was the nature of the shed
culture that a fireman soon got a reputation, good or bad, and this was
important for it could directly affect the career of the individual.
Walter Mulligan describes the process:

> After they got used to you, you were treat as an equal more or less
> you know, they still wouldn't lift the hand lamp off but they would
> talk to you. You weren't frightened of them but you were in awe of
> them you know. What they said was law, but when you got firing
> for them regular and they knew you could do the job they had no
> bother with you.
> (Walter Mulligan, interviewed 1995)

Later on in the fireman's career his reputation would be important if
the driver was to allow the younger man to take the controls of the
locomotive for an unofficial drive, an important way in which firemen
learnt the craft of the footplateman. Training was empirical, with the
fireman observing his driver, whether good or bad, at work, or being
shown good practice by the senior man pointing out particular techniques or the peculiarities of the road they found themselves on.[5] For
the keen fireman there was much to learn, and a conscientious driver
would be imparting knowledge all the time. The reason why this learning process took the form it did was because of the complexity within
which the simple function of raising steam took place. These variable
conditions included the way the engine was being driven, there being
an enormous difference between the most efficient and inefficient
combination of practices. Here Frank Mason, who worked on the LMS

before nationalisation, describes the art of creating the perfect fire in the locomotive:

> The fireman is especially alert in these early stages. He needs to assess the quality of the engine, the fuel, the load and the driver; if not his regular mate or known by reputation. He closes the door and watches the chimney for smoke. If it is black he will open the door a little, smile and sit down, or prepare coal for his next firing. If the smoke is patchy he takes hold of the shovel, opens the door and using his shovel blade as a deflector of air, directs the air in a sweep across the fire to see if one part is brighter than the whole. Without this trick of experience, the tremendous glare and heat prevents good observation.
>
> (Mason, 1992: 23)

Part of the training was to learn the road they were travelling on and the controls of the engine. Here again the driver would, if he was willing, act as a mine of knowledge, drawing the younger man's attention not only to the everyday but also to the extraordinary that might only crop up once in a working life. Sometimes incidents that had happened to others in the shed would be used as illustration, thus building up a reserve content of knowledge and skill. Joe Brown, originally of West Auckland shed, and Tom Storey of Pelaw shed, interviewed in 1995, discuss the nature of training:

> *Joe Brown*: The drivers trained you or you trained yourself, there was no [official] training, you just learnt off the driver and he would tell you. They were very good like, some of the old drivers.
>
> *Tom Storey*: You know, brilliant, and they knew everything. You used to go to work one day and they would be asking you all the questions, steam actions and all this kind of thing.

When the fireman was near 'his time', before being passed for firing or driving duties, attendance at the Mutual Improvement Class (MIC) was expected. These classes were held and run by the grade themselves to ensure younger members successfully passed the formal examination by the footplate inspector. The MICs included both theoretical and practical aspects of the job and were often held in halls owned by the railway companies. Here the men would be given lectures and practical demonstrations of the various aspects of their craft, including the handling of sectionalised parts of steam locomotives that had been donated or 'acquired' by the local shed. Interested drivers, and sometimes inspectors, took the classes. Attendance at the MIC was entirely voluntary, and unpaid, both for the teachers and the taught. Here Tom Storey describes the sacrifices he and his family made:

> I travelled 12 miles to the MIC ... In fact on a Sunday morning I used to go to Consett MIC and it was in your own time. I used to get the first bus, maybe 8 o'clock in the morning and it used to be 3 o'clock and 4 o'clock before you were coming home ... I think I went five year or something like that.
>
> (Tom Storey, interviewed 1995)

Underlying this sacrifice was the fact that if the worker failed to attend, his chance of passing the formal examination would be lower. As Tom Storey explained:

> You knew you had to; it was as simple as that. I'll tell you what, any boy that didn't get their selves involved used to have to work harder at it when the time came. There were any amount of them that didn't pass just through not going to the classes ... when you were passing to be a fireman, passing to be a driver, things like that, you had to do it because you couldn't look at books and do it.

A report from the secretary of Hull LNER MIC in a staff magazine of 1923 gives an account of the kind of subject matter covered in such classes:

> Lubrication, the Westinghouse Brake, Superheat, the Vacuum Brake, the Control of the Valve from the Footplate and Rules and Regulations for Enginemen.
>
> (NER Magazine, 1923, Vol. 13, Issue 152: 220)

The article goes on to note that the branch had a membership of 120, with weekly attendance of between seventy and eighty drivers and firemen. In addition to the MIC, the footplate workers' union ASLEF produced an immense amount of detailed advice in the columns of the *Locomotive Journal*, where members of the grade would debate highly technical topics and pass on accumulated knowledge.

This socialisation system, which combined technical, social and moral aspects, is reflected elsewhere in accounts of railway life. In the signalling grade there was some limited 'theoretical' training given by railway companies, but the prospective signallers would learn their trade from those already practising it. This was a very direct way of being socialised into both a work and occupational identity. Here Bradshaw (1993: 25) describes what it meant for him:

> At fifteen and a half years of age, I was quickly to learn the meaning of maturity and manhood, for here I was a lone teenager thrown into a world of adult working men, without a single person of my own age group for companionship or consolation.

Bradshaw also describes the process of this socialisation. Here he discusses being taught the role of booking lad during the 1940s:

> By the eighth day that hitherto impenetrable barrier had been conquered and Ted Cox's face broke down into a satisfied smile. With a pat on the back he announced 'You'll make it lad. Now we'll show you how to write. Your script is appalling.' ... Up to then, I had secretly feared him; now I felt a conversion to almost hero worship.
>
> (Bradshaw, 1993: 21)

Importantly this form of socialisation was not simply empirical in nature, but also included the transmission of a great deal of theoretical and technical knowledge. The formal training in a signal box might only consist of spending a week on each of the shifts. This would ensure that the trainee had experienced a range of traffics and situations that arose with a particular timetable. If trainees were fortunate, senior staff would also pass on other knowledge about the box and the area it controlled that would not come up during the period of training, and indeed might never arise in the working life of most workers in the grade. Interestingly, this kind of advice and information was not limited to the area – workers would also draw on examples of incidents on other systems and companies to illustrate a point, and again this built up tacit knowledge in the new worker. Ron Bradshaw describes the aftermath of his first mistake in the signal box in front of a qualified signalman. Bradshaw, having set the road for an express train, had not noticed that the signal had failed to clear.[6] The driver of the train had been given misleading information and had to brake dramatically in order to stop:

> Joe [*the qualified signalman*], rather than chastise me after the visible shaking-up, was feeling sympathy but repeated the importance of

observing a constant vigilance and the importance of signal arm response and track circuit indicators. He quoted the Castlecary (Scotland) accident of 18 months ago when 35 people were killed. During the ensuing enquiry a signalman was held partially to blame for incorrect observation of the indicator of a distant signal. This little episode had a sobering effect on my possible over self-confidence and was to remain a constant reminder when working busy boxes later, a truly graphic illustration of 'more haste, less speed'!

(Bradshaw, 1993: 29–30)

Both the footplate and the signalling grades enjoyed similar forms of socialisation and training, in the sense that workplace knowledge was largely the autonomous product of the grades and wider workforce. This system became embedded in the very structure of the industry as it developed, partly because of company need, but also as a result of the sheer stability of railway work over time. Knowledge was built up over a working life by individuals who drew on, and eventually contributed to, a collective workplace tradition. The process of socialisation took place within the context of embedded historical knowledge about the industry. For example, many of the older and retired workers I interviewed from the footplate and signalling grades in the mid- to late 1990s had enjoyed a career of up to 50 years' service, joining the industry just before or during World War II. They had learnt their craft from workers who had themselves worked on the railway for up to half a century or more, taking the historical experience of working life back to the turn of the twentieth century or before. This is not to say that all workers spent their lifetime in railway service, or were always trained by someone quite so senior. But it is apparent that there was a sense of history embedded within the job of many workers, whose collective memory was shaped by their own long-term experience and that of others. Susan Faludi (1999: 73) touches on a similar aspect of shared work culture in her discussion of the role of shipyard 'fathers' in the context of US naval repair yards:

To be a shipyard 'father' in the later years was to have command not over men but over a body of knowledge – and to be capable of transmitting that knowledge to a younger man who would, through his mastery, become a teacher himself. The more knowledgeable man was the 'father' not simply because he had authority but because he was willing and able to confer some of that authority upon another.

Stories, memories and skill would, therefore, become removed from their original source and become an active element in a live historical collective process. Raphael Samuel has talked about such a phenomenon as the 'theatres of memory'. Here he describes his thesis:

> It is the argument of *Theatres of Memory*, as it is of a great deal of contemporary ethnography, that memory, so far from being merely a passive receptacle or storage system, an image bank of the past, is rather an active, shaping force; that it is dynamic ... and that it is dialectically related to historical thought, rather than being some kind of negative other to it.
>
> (Samuel, 1994: ix–x)

He goes on to argue for:

> the idea of history as an organic form of knowledge, and one whose sources are promiscuous, drawing not only on real-life experience but also memory and myth, fantasy and desire; not only the chronological past of the documentary record but also the timeless one of 'tradition'.
>
> (Samuel, 1994: x)

What I want to suggest here is that the railway industry is a powerful example of agency as historically reflexive. History acted as a knowledge bank or resource which railway workers in general were, and still are, able to draw on for meaning and understanding. Such a system also creates a very real sense of ownership of, and identification with, work itself. Such an occupational identity represents both technical as well as moral control over the labour process.

Leviathan challenged

Victorian railway management could be described as autocratic paternalism, with workers of all grades in highly dependent relationships with their employers. The contradiction for paternalist organisations such as the railway companies was that the measure of stability which parts of the workforce enjoyed, when coupled with authoritarian labour policies, had the effect of stimulating the growth of an organised opposition. Early attempts at organised opposition had usually taken place in an ad hoc manner by specific grades in particular companies. Individuals or groups of workers could petition directors, and Bagwell (1963) notes that companies made much of the access that was afforded by them. Employers argued that there was no legitimate need for intercession between company and individual servant. The inequality in the employment relationship was reflected in the humble language adopted by these petitions and is often remarked on in trade union histories written for the members. The semi-feudal character of early railway work is contrasted with the later equality won in the face of trenchant opposition from the companies (Raynes, 1921; Alcock, 1922; McKillop, 1950).

In 1871 the first permanent union in the sector was formed. The Amalgamated Society of Railway Servants (ASRS) was a general union which aimed to represent all workers in the industry. In reality, only the more privileged staff were able to join – signalmen, footplatemen and guards. It was founded with substantial backing from the middle classes, most notably Michael Thomas Bass, the Liberal MP for Derby. This early support left a heavy imprint on the Amalgamated's industrial relations policy, which was conciliatory in tone, eschewing confrontation with employers in favour of negotiation. The ASRS, like other Victorian unions, sought to build a rational case for collective representation, and traded on the image of respectability and reliability that the railway worker could command in public opinion. The appeal of liberal politics was based on the conviction of the fairness of railway labour's case and the intransigence of the employers. Increasingly unpopular, the companies were often portrayed as greedy monopolists reluctant to invest in the latest safety devices, as unconcerned with passenger safety as they were with that of their staff. Jack Simmons' (1994) essay on the *Punch* cartoonist John Tenniel, whose work included many lampoons of railway management (see Figure 2.3), recounts the way the satirical periodical waged a consistent campaign against company mismanagement, in particular their parsimonious attitude to investing in safety equipment while being content to let staff take the blame in the aftermath of accidents. As Simmons points out, *Punch* reflected the views of its readership rather than leading opinion (see also Figure 2.4). Public sympathy for company servants buttressed the case for company recognition of trade unions, which was only conceded in the case of the North Eastern Railway at the end of the nineteenth century (Howell, 1999).

It was frustration with the moderation of the ASRS that led to the creation of the Associated Society of Locomotive Engineers and

RAILWAY RESPONSIBILITY.

Mr. Punch. "NO, NO, MR. DIRECTOR, *THEY'RE* NOT SO MUCH TO BLAME. IT'S YOUR PRECIOUS FALSE ECONOMY, UNPUNCTUALITY, AND GENERAL WANT OF SYSTEM THAT DOES ALL THE MISCHIEF."

Figure 2.3   Railway Responsibility by John Tenniel. (*Punch*, 1874, Vol. 67, p. 129)

Firemen in 1880. This sectional craft union drew its membership purely from the footplate grades, and their skilled status allowed a far more confrontational stance towards management to be taken. From its foundation ASLEF differentiated itself from its rivals, particularly the ASRS, by way of an appeal to craft pride and the sense of 'calling' of its membership. The histories of the footplate union, commissioned by ASLEF itself, attest to an Enlightenment faith in progress, with sections in both Raynes' *Engines and Men* of 1921 and Norman

THE PATENT SAFETY RAILWAY BUFFER.

Figure 2.4   The Patent Safety Railway Buffer, Anon. (*Punch,* 1857, Vol. 33, p. 25)

McKillop's *The Lighted Flame* of 1950 charting the 'forward march of labour'. As Baty, general secretary of ASLEF, wrote in his foreword to McKillop's book:

> Consider our calling in the days of the Society's founders, when men miraculously preserved pride of craft despite appalling hours of work performed for a pittance in conditions of semi-slavery; and then read of the landmarks that one by one have been reached and passed by valiant effort and struggle.
>
> (McKillop, 1950: vi)

The paragraph continues with a list of events and battles won that reads like a regimental history: hours of work, recognition, membership, funds and holidays. Baty goes on to stress that this achievement was realised by the union drawing on its social resources:

> I may be pardoned if I briefly point the moral that I believe it [*the Union's history*] conveys. It is, I think, that our Society owes its outstanding record of achievement to the fact that at no time has it functioned merely as a machine; it has been also at all times a living brotherhood of men prepared to devote themselves to the uplifting of our great craft and to the general welfare of all engaged in it.
>
> (McKillop, 1950: vii; original emphasis)

ASLEF and ASRS were joined by smaller unions representing other elements of the railway workforce, including the Railway Clerks' Association in 1897, the General Union of Railway Workers in 1889 and the United Pointsmen's and Signalmen's Society. The unions won limited acceptance from the employers in 1907, with the creation of the railway conciliation scheme. Further concessions were granted after the Great War. While the interwar period was not without setbacks for trade unionism in the industry, particularly after the General Strike of 1926, organised labour enjoyed general recognition. Both employer and employee worked within a complex system of workplace law and agreements which greatly restricted the arbitrary power of management.

A railway culture?

So in what sense can we talk of a distinct 'culture' being established in the railway industry in the nineteenth century? Railway employment has perhaps suffered from stereotyping, with accounts tending to compress and condense individuality, producing at times a deformed caricature of work identity. As Savage (1998) highlights, there is a tendency in railway historiography to imply that career structures developed to the point where they offered a common experience to workers across the system from very early on. This produces a rather static account of both the industry and its workers, which is reflected not only in the nostalgia that surrounds the railways but also in academic analysis. For example, in his discussion of stable occupations, Hollowell (1975) talks of the 'remarkably constant fortunes of the British Railways locomotiveman'. The result is a creation of ideal types,

Figure 2.5 Corporate pride: *The Permanent Way – Relaying* by Stanhope Forbes. Promotional poster for the London Midland and Scottish Railway, 1924. (National Railway Museum/ Science & Society Picture Library)

stereotypes and of a fixed notion of occupational identity in a highly ahistorical way. This kind of stereotyping could never capture the full range of identities that existed in the railway workplace, nor the processes by which they were created.

Railway culture, or to be more accurate, *cultures*, were born out of a peculiar mix of forces. Railway companies were themselves distinctly modern in their bureaucratic structure and centralisation of command and control. Yet simultaneously they were paternal, even feudal, in their relationship with employees. Added to this cocktail was the strong military flavour found in early management. The companies needed reliable, disciplined, dedicated and skilled workers, and could offer security and promotion prospects virtually unknown for working-class men in the Victorian economy. But the prospect of a job for life, arguably the archetypal example of such security, was conceded at the cost of tremendous dependency.

The nineteenth-century railway company worked hard to project a distinct corporate identity. Each of the companies resembled its own State, each had its own territory and boundaries, its own livery and coat of arms, its workforce – the railway army. Yet within these institutions workers formed identities based not only on their companies, but also on a range of other differences such as department, station, shed, region, grade and craft/non-skilled status and trade union. What united these disparate identities was the work undertaken and the commonalities by which successive generations had learnt how to be railway men or women. If there is a definition of 'railway culture' in this book, it is to be found in the characteristics of autonomy of work, training and socialisation in the industry. It is in the way knowledge about work and the industry was stored and transferred between and among workers. And, finally, it is in the way this culture has been inherently dynamic and adaptable over time.

Culture is, as many writers have pointed out, a notoriously difficult concept to define in the social sciences (see, for example, Williams, 1983; Eagleton, 2000; Smith, 2000). Culture can be thought of as a way of life of a society or of a part of it. It implies a shared set of values and norms, ways of understanding and interpreting the world. Culture, and especially the term 'railway culture', will often crop up in the rest of this book. In what follows I am not trying to impose a reading or definition of culture on a set of workers in the industry; rather I am interested in the way the term culture is used by various groups, including politicians and management writers, managers and trade unionists in the industry, and journalists. In this sense I am following Mark Smith (2000: 20) when he says:

> In short, the concept of culture is an open window through which we can identify the assumptions, values and classification systems at work in a particular location.

Smith goes on to acknowledge that the way in which culture is defined tells us a great deal about the approaches adopted by writers and thinkers. It seems to me that this fascination with culture – what is 'right', but more often what is or was 'wrong' with it – tells us as much about our own culture as it does about that of the workers studied here.

The following two chapters examine the way the rail industry was shaped during the period of nationalisation up until 1979. Chapter 3 deals with the management and organisation of the railways, while Chapter 4 discusses the nature of work at this time.

3
On Behalf of the People: Managing a Nationalised Railway from Attlee to Thatcher, 1948–79

> If the full advantages of co-ordinated transport are to be secured we must eliminate the conflicting interests created by the existence of innumerable separate private ownerships. These, quite naturally, are chiefly concerned about advantages for their own individual undertakings. They do not worry themselves too much about the efficiency and welfare of transport as a whole. If this is agreed, we then have to choose between private monopoly and public monopoly. The trouble about private monopoly is that it is dangerous for the general wellbeing. The management of a private monopoly must inevitably be fettered by much state regulation to protect the interests of the community. A monopoly which is publicly owned does not threaten public interests and so does not require such intense state regulation.
>
> (Herbert Morrison, *British Transport at Britain's Service*, 1938, quoted in Bonavia, 1971: 40)

Introduction

Over one hundred and twenty years of private ownership came to an end with the nationalisation of the railways on 1 January 1948. One could be forgiven for thinking that this marked a sea change in the way the industry was operated and managed, but to what extent did fundamental change really occur? This chapter examines the problems and contradictions for management in the railway industry over three decades ending with the election of the first Thatcher administration in 1979.

Nationalisation represented a change of legal ownership but was marked by continuity in organisational and managerial structures, and

43

this had important implications for both workers and managers in the industry, even shaping the era of privatisation described in later chapters. Management grades continued to be embedded within the industry, part of the culture of the organisation itself, and acted as a stabilising influence during a period marked by a series of reforms and rationalisation as historic commercial and economic problems faced by the railways were finally tackled. But this 'railroad culture' came under sustained attack during the period from those who saw it as part of the problem the industry faced. A particular criticism was that railway management was too narrow in its focus, being driven by sentiment rather than commercial strategy. Pivotal in this story is the figure of Richard Beeching who, as an outsider, was deliberately appointed to the chair of the British Railways Board (BRB) to tackle inefficiency within the industry.

Nationalisation 1948

As part of the postwar Labour Government's commitment to public ownership of certain key industries, the railway industry's turn to be nationalised arrived during the third year of Attlee's administration. The four large 'Grouping' companies of the interwar period – the LNER, LMS, SR and the GWR – now became British Railways, with the Railway Executive (RE) at its head. This executive was itself part of a larger organisation, the British Transport Commission (BTC). The BTC was effectively the holding company for all of the nationalised inland transport undertakings which, like the railways, had entered public ownership from the beginning of 1948. The RE was the largest of five executives under the umbrella of the BTC, the others being for London Transport, Docks and Inland Waterways, Road Transport and finally Hotels. The BTC itself reported to the Minister of Transport. Below the RE was a series of six area boards: Eastern, North Eastern, London Midland, Scottish, Southern and Western, roughly based on the former companies (Bonavia, 1971: pt. 2). The BTC, and therefore each of the executives, was given the rather broad and vague objective 'to provide ... an efficient, adequate, economical and properly integrated system of public inland transport', with the aim of breaking even 'taking one year with another' (Pollins, 1971: 168). The nationalised sector, therefore, was never set the task of profit maximisation, but rather a far more ambiguous objective of serving the public good.

Calls to nationalise the railway industry had been made in trade union and radical circles since before the turn of the century, and as we saw above the State had allowed itself, under Gladstone's 1844 Railway Act, the ability to take control of the industry (Foreman-Peck and Millward, 1994; Crompton, 1995). The Labour Party had made a broad commitment to public ownership in its adoption of Clause IV, as part of the revision of its constitution in 1918 (Pelling, 1976: 43–4). When the decision to nationalise was taken, it was based on several factors, not simply on this pre-existing, and rather vague intention embodied in Clause IV. The policy emerged, firstly, out of the experience and outcome of the war, with its legacy of destruction, backlog of repairs and renewals. Secondly, there was the growing enthusiasm for 'planning' in many branches of the economy, and in society more widely, to ensure efficiency and productivity. As Cunningham (1993: 3) writes:

> In the decade before the Second World War, the idea of some form of central economic planning was at the heart of the Labour Party's economic strategy.

And in the context of the railway industry, David Howell (1999: 345) suggests:

> The post-war Labour Government was intensely serious about the necessity for industrial modernisation ... the railway industry, with its inter-war difficulties and the wartime depletion of its assets, appeared a prime candidate for this agenda.

Transition to State ownership did not entail the wholesale removal of either directors or personnel from the formerly private industry. Gourvish (1986: 35–7) discusses this continuity in the Railway Executive:

> The Railway Executive of 1948 was essentially a body of experienced railwaymen of the old school. Their average age on appointment was 57. General Sir William Slim ... was the only full-time member with no prior knowledge of railways. The others had all joined the industry before the age of 21 and had accumulated an average of 40 years' service by 1948. With the exception of Michael Barrington-Ward and John Landale Train ... all of the group had entered the industry straight from school and had thereafter worked for only one company.

And the only concessions in the direction of workers' control were in the appointment of W.P. Allen, formally general secretary of ASLEF, onto the board of the Railway Executive, and of John Benstead, who held the same post in the NUR, to the board of the BTC. Both these men resigned from their respective union posts to guard against conflicts of interest, again showing the commitment to consensus (Gourvish, 1986: 123–4). But as Howell (1999: 353) points out:

> Both these former officials had been impeccably moderate during their trade union careers. In no sense could their new posts be seen as an injection of radicalism into the managerial hierarchy. Union criticism of the industry's organisation was concerned with the lack of scope for workers' participation in significant decision making.

There was, however, a genuine commitment on the part of the BTC, along with other nationalised undertakings, to be seen as good or even model employers (Pendleton and Winterton, 1993: 1–2).

The employment of so many managers from the former private companies is easy to appreciate in the circumstances of postwar Britain but it entailed, as was the case in other industries, the allocation of senior posts to those who had been and still were, actively hostile to the concept of public ownership. A.J. Pearson, who had been an officer in the LMS and later went on to become a member of the BTC, describes in his autobiography drawing up and mounting a campaign of opposition to plans to nationalise the industry (Pearson, 1967: ch. 7). John Elliot's memoirs provide a similar account, and show that the motives for the campaign were somewhat more material than ideological:

> It was decided, rightly in my opinion, to launch a public campaign against nationalisation, the real objective of which was to obtain the best compensation for the Companies' shareholders who, it was felt, would be at risk if the Boards agreed to nationalisation without a struggle.
>
> (Elliot, 1982: 66)

These same officers and board members of the old Grouping companies were the very people who now populated the new Railway Executive. Elliot tells of his misgivings about the new organisation and that these were shared by many colleagues including Eustace Missenden, former Southern Railway general manager and first Chairman of the RE:

> Missenden told me privately that he heartily disliked the whole thing [nationalisation and its organisational structure]. As Chairman,

he was merely 'primus inter pares' with his colleagues on the Executive, and in no sense Chief General Manager of British Railways. He added that he had made his mind up to retire as soon as possible ... The whole thing was bound to break down sooner or later, and I should try not to get involved.

(Elliot, 1982: 68)

Elliot, who succeeded Missenden as General Manager at the Southern, was also offered the chairmanship of the RE on Missenden's retirement in 1951. He was surprised by the offer from a Labour Government, admitting to Alfred Barnes, the Minster of Transport:

I am not a Socialist. In fact I was a member of the Carlton Club until 1948 and resigned only because I didn't think it was right to accept a major post in a nationalised industry and remain a member of a Conservative club.

(Elliot, 1982: 80)

It is clear from the rest of Elliot's reflections on these events that neither Barnes nor Attlee saw his personal politics as any barrier to his appointment. There was, therefore, at the top of the RE a group of men who had actively opposed State ownership, who believed that the form it had taken was flawed, and even doubted its long-term survival. As we have seen, the decision to ensure a degree of continuity was based on the pragmatism and specificity of the period, but it also has to be placed in the political context of the time. Within sections of the Labour movement there was hostility to the idea of workers' control. As Cunningham (1993: 3) has argued, 'support for "bottom up" forms of economic organisation, for example Guild Socialism or more militant syndicalism, was on the wane'. This can be explained partly by the idea that such control would conflict with the aims of a centrally planned efficient economy, but it also reflects the elitist fear of the 'workers' within the Labour Party itself and, more broadly, amongst part of the electorate.

Distrust of the 'workers' and, by extension, of workers' control, can be seen as a product of the class background of some of the leading intellectuals in the Labour Party. In his book *The Intellectuals and the Masses,* John Carey (1992) charts suspicion by the Webbs among others of a more democratic extension of industrial decision-making. Attlee's administration aimed at consensus, rather than at representing particular interest groups, in spite of the rhetoric of 'being the masters now'.

Ross McKibbin (1991: 5), in his writing on the later Wilson and Callaghan administrations, talks about the way 'Labour always plays the game by other people's rules', and in many ways this is prefigured in Attlee's policy. Therefore *ownership* of the industry, rather than the way it was run, was seen to be the crucial principle.

The financial arrangements of nationalisation gave even greater cause for pessimism. The main problem for both the BTC and the RE was the arrangement by which the shareholders of the former companies had been compensated. In order to effect a smooth transition to the public sector, the shareholders were given generous compensation, far above what their assets were worth (Gourvish, 1986). The successor organisations had to service the interest payments on the debt this created, and their problems were compounded still further by the financial aims and objectives set for them by the Government. In short, the railway was expected to provide a public service at as cheap a rate as possible. It achieved this by the cross-subsidy of services, thereby not covering the costs on many of them. Partly because of the public-service objective, the RE failed to build up sufficient funds to renew assets and pay its share of the inherited interest charges. This had the effect of forcing the RE, and all the public corporations, into the more expensive strategy of borrowing capital for investment rather than financing it out of profits, so increasing debt and adding still further to future interest payments. This situation was made worse later on by successive postwar Governments using the nationalised industries as blunt instruments of economic control, and the downturn in traffic levels (Thompson and Hunter, 1973; Saville, 1993; Foreman-Peck and Millward, 1994).

The experience of management

The continuity in senior personnel was reflected at lower levels in the new British Railways. For many managers and supervisors there was little initial change with the onset of public ownership. Frank Hick, who began his career on the North Eastern Railway in 1922 and was later to become operating chief of the North Eastern Region, reflects on the transition:

> Strangely, this did not make much impression on me and did not change in any way my outlook towards my work ... and that is how many of my contemporaries felt.

> (Hick, 1991: 74)

Likewise Frank Ferneyhough, who had been a supervisor before nationalisation, noted his feelings:

> Most of us in the railways wondered anxiously what would happen. Nothing did, at least not immediately of any significance. The trains still ran. Locomotives still chuffed, and ticket collectors clipped the tickets. Changes could come only slowly, and the immediate impact was sure to be far greater in the management structure than at stations, depots and works.
>
> (Ferneyhough, 1983: 153)

Of necessity, lower levels of management and supervisors were indispensable to the safe and efficient operation of the railway, and were deeply embedded in the structure of the industry itself. These grades seem to have viewed themselves not as generic managers, but rather as fundamentally railway workers in much the same way as drivers, guards or signal workers did. In part this situation had emerged with the development of the industry in the nineteenth century. While railway management had to be recruited initially from outside the industry, often from the armed forces, there quickly developed a sophisticated organisational hierarchy with its own internal labour markets. Potential managers were drawn from the lowest levels of the white-collar sections of companies. Although patterns varied, at their most formal companies operated their own training scheme for management. The NER, for example, instigated its own traffic apprentice system before the turn of the century (Bell, 1951; Irving, 1976). Such was the success of this method of developing future managers that it was continued by the LNER and formed the basis of BR management training after 1948, and was only abandoned during the lead-up to privatisation.

Training aimed to give the would-be manager a range of railway operating and administrative experience, to expose the individual to a series of unique situations throughout a system. This more formal scheme valued a hands-on approach to training, with traffic apprentices moving round a company's system, staying in a location perhaps no longer than six months at a time. Gerard Fiennes (1973), who was recruited by the LNER traffic apprentice scheme in the late 1920s, gives a glimpse of the range of experiences he had during his time on the scheme, from learning how to build the brick arches in the fire box of a locomotive, to the putting of drunks on trains. He stressed the ways learning on the job was valued:

> We spoke to an officer perhaps ten times a year. We were taught our trade by the men who plied those trades.
>
> (Fiennes, 1973: 1)

Many of the senior managers in British Railways who had spent their whole career on the railway were a product of this scheme, which relied heavily on trainees doing 'real jobs' on the ground. This stress on experience rather than theory was also the preferred method by which those managers who were promoted through the ranks were also trained (Pearson, 1967; Ferneyhough, 1983; Hick, 1991).

Common to both career ladders was the way in which the individual concerned was socialised, or embedded, in the culture of the railways, becoming a 'railwayman'. Most of the autobiographies considered here reflect at some point on the idea of 'being a railwayman', and it is obviously done with a sense of pride. The important point to make is the access that such accounts allow us in mapping the relationships within different management levels as well as between management, supervisors and workers. What was created over the length of a career was a mutual respect between management and workers over some aspects of work organisation. There were of course divisions between such groups as a result of their contradictory positions in the employment relationship (Brown, 1988), but it is important to understand how the employment relationship at the point of production is mediated, humanised and made meaningful. It is not experienced usually, or even occasionally, as the direct clash between capital and labour, but rather is always embedded in real personalities at specific points.

In the railway industry, the employment relationship and the respective identities of workers *and* management were the products of generations of socialisation. In some cases the sort of early or anticipatory socialisation into the industry for managers and supervisors was similar to that experienced by many railway workers. Awareness of, and enthusiasm for, the railways was an important part of their decision to pursue a career in the industry. Frank Ferneyhough describes his early experience 'brainwashed by the vivid sights and sounds of steam railways' (1983: 20). This particular account is valuable for the sense in which this interest in, and respect for, workers in the industry stays with Ferneyhough throughout his career as a supervisor and manager. Here he describes the awe in which he held drivers, workers he was in charge of as a station master:

> Engine drivers can be formidable fellows. When necessary, they can cut you down to size by doing nothing – just ignoring your very existence in a deep, dark grey, stony silence. For me they were a very special breed. Not like ordinary chaps.
>
> (Ferneyhough, 1983: 87)

Such an attitude also stresses the importance of situating these encounters within a framework that recognises the centrality of the stage the individual has reached in his life cycle. Ferneyhough was appointed to his post while he was in his mid-twenties, while the grade he is so obviously in awe of here contained workers who could be in their mid-forties, fifties or sixties. Ferneyhough's relationship with such men is permeated by respect for their grade and experience, as well as by an emotion that would be recognised outside industry: simple respect for age. There was a moral quality to the intersubjective relationships between grades and management and supervisors. This element of respect is also reflected in other autobiographies, notably that of Hick, who describes his thoughts about the members of the relief signalling grade he managed in the York area during World War II:

> Their knowledge of the signalling and safety regulations, at that time in many different boxes, and their ability to work the Morse single needle telegraph instrument were fundamental to the job, and I constantly marvelled at their contribution to the running of the railway ... A highly intelligent set of men, they needed careful handling, but once you got their confidence they proved to be the most loyal and supportive bunch of fellows one could ever wish to meet.
>
> (Hick, 1991: 56, 59)

It is clear from many such accounts of managerial life in the industry that much store was put in being a 'railwayman', i.e. developing an attitude to the work that valued the specificity of the railway industry above that of being a 'manager'. Elliot, who had enjoyed a military background before joining the Southern Railway, speaks here of his 'conversion' after joining the operating department:

> So it was that by degrees, almost sub-consciously, I found myself thinking as a railwayman. How I did it I don't really know, but of

this I am sure, that without the patient instruction of those older and far senior to me, I should have been sunk without trace.

(Elliot, 1982: 45)

In Fiennes' book we get a series of descriptions of his respect for railwaymen, and clearly he considers himself one of them – one of his own self-imposed rules being 'Behave to railwaymen as a railwayman' (1973: 30). In the same volume Fiennes discusses his relationship to his work:

The story is about someone who came slowly and reluctantly to grips with the basic realities of a railway job and against all expectations and reason was absorbed by the magic of the thing.

(Fiennes, 1973: vii)

Nowhere is the quality of the embeddedness of management in the industry reflected more vividly than in the context of war. In all the books written by those who experienced World War II it is clear that the events of that period had a profound and lasting effect on their behaviour and attitudes towards those that they were, or would later be, in charge of. Here Hick (1991: 55) pays tribute to railway workers in the war:

I can only say again and again what valiant chaps those trainmen were, as indeed were all railwaymen working at ground level during those hazardous and uncomfortable days.

He describes the aftermath of the bombing raid on York in April 1942, which saw the railway targeted and the station suffer a number of direct hits:

Soon we had to help the staff from the station who knew of our hide-out and who were in various stages of injury and shock. My friend John Grant appeared, dishevelled, battered and bruised, black and grim, unable to understand how he had escaped from a totally collapsed building ... Everyone who could help rallied round, and outside Inspectors acted as additional support in whatever direction they were needed ... a philosophical mentality developed, and people just 'had to get on with it'.

(Hick, 1991: 61–2)

Gerard Fiennes, who had also personally experienced the attack described above, reflects here on the experience of war in teaching him his trade:

> railways and railwaymen have a great scope and a great gift for improvisation ... we tend to forget that in 1939 and 1940 we made radical changes to the freight working through the controls at an hour or two's notice, and that we followed it up with a Train Circular next day. So I learnt how to improvise, how to clear or get round – mentally and physically – accidents, traffic jams in yards, terminals and lines. I have never been frightened of any traffic situation since.
>
> (Fiennes, 1973: 25)

The war had the effect of ontological or existential levelling within this industry, and perhaps many others. Essentially what occurred was the revealing of a common humanity within extreme conditions, which at times threw up heroes, but more usually simply disturbed hierarchies. This experience of war and the respect it engendered could, however, be double-edged, with the work practices essential to wartime difficult to combat in the postwar period. Hardy describes the tension that management faced, in his case seven years after the end of hostilities:

> in many large passenger and freight terminals, marshalling yards and locomotive depots, practices grew up, again as a matter of expedience, that would never have been tolerated before the war and should never have been allowed to persist when peace returned. Only too well do I know the patience, determination and, dare I say it, the degree of courage needed to eliminate the bribery and corruption that had been used to get work done under the dreadful wartime conditions in London, particularly during the bombardments.
>
> (Hardy, 1989: 19)

Management in the postwar period bore a heavy legacy of tradition mixed with the experience of war. What had developed was a deep sense of the management within the industry being part of it, not a discrete practice undertaken by a separate group. This is not to say that the industry was closed to new ideas or that training was completely empirical. The traffic apprentice scheme, for example, was seen as an excellent example of management pedagogy. Rather, railway managers were embedded in work by the processes of socialisation, training and managerial practice. What the memoirs from wartime amplify is the moral aspect of these social relationships, whereby industrial relations were overlain with personal ones. Hardy's (1989) account illustrates the very real problems that this presented for management in the postwar period. It was difficult to tackle the custom and practice that had taken root in wartime because the actors on both sides of the employment relationship were implicated in them.

It must also be acknowledged that much of the possible tension between capital and labour was defused or redirected by the elaborate form of industrial relations that had been a feature of the industry since the first decade of the century. This had grown into a multi-layered procedure with scope for discussion at local area, regional and national levels. This system was itself coupled with a deeply embedded set of rules and regulations that provided great stability in the industry during the period (see below, and also Gourvish, 1986: 120–33; Ferner, 1988; Pendleton, 1991a, 1991b; Edwards and Whitston, 1994).

Changing directions: 1951–61

With the return to power of the Conservatives in 1951, the structure of the industry was changed again. The road transport element of the BTC empire – an area that had been heavily invested in by the State – was privatised, and the Railway Executive was abolished, the BTC thereafter taking direct control of the railways centrally. The aim was to remove unnecessary layers of bureaucracy while devolving power to the regional level, but in reality it created even greater levels of central-isation (Bagwell, 1982; Gourvish, 1986). In part this was due to the appointment of Sir Brian Robinson (a former general in the British army) to the post of Chairman of the BTC. He introduced a military-style chain of command, further removing power from lower levels of the organisation, and concentrating it in the centre.

A long-awaited investment programme in the industry began in 1955. The Modernisation Plan cost £1,240 million, and aimed at the complete renewal of the rolling stock and much of the infrastructure of the industry over a period of 15 years (Bonavia, 1981; Gourvish, 1986). The plan was to see the complete eradication of steam traction from the railway, to be replaced by electric or diesel motive power. This investment is viewed by most commentators as having been wasted on ill-thought out and untried designs, and on projects that were decades too late (Fiennes, 1973). In essence, the Modernisation Plan attempted to replace, like for like, what existed before the plan, the strategy being that if the industry could update rolling stock and other facilities, it could compete effectively with the road transport industry and the private car.

Gourvish (1986) has attempted to be balanced in his analysis of the Plan and the investment that flowed from it. He points to the kinds of pressures exerted on the industry and its management. British Railways was still attempting to provide a national public service at low cost, while simultaneously being under intense political pressure pushing it in a contradictory direction. On the one hand, there was an expectation of the business breaking even and investing wisely, and on the other, criticism if any attempt was made to rationalise or close parts of the system. As Gourvish (1986: 209) writes:

> The social service aspect of railway marketing remained firmly in the minds of most senior railway officers, and it was encouraged by the public clamour whenever closure proposals were announced.

Such a clamour is reflected in various aspects of popular culture, and predates the widespread closure programme of the Beeching era discussed below. The prospect of a clash between Whitehall bureaucrats and village communities attempting to prevent the closure of an unprofitable line was too great a temptation for Ealing Studios, which caricatured such a battle in the 1952 feature film *The Titfield Thunderbolt*. In the film, a group of villagers lead an attempt to save the branch line from closure by offering to run the service themselves. During the public enquiry into the closure, it appears that the group has lost its chance. This elicits an outburst from the local squire:

> Don't you realise you are condemning our village to death? Open it up to buses and lorries and what's it going to be like in five years' time? Our lanes will be concrete roads, our houses will have numbers instead of names, there will be traffic lights and Zebra crossings ... [the railway] means everything to our village.

While Titfield was celluloid invention, it was nonetheless based on the real story of the first preserved railway in Britain, the Talyllyn line in Wales. In his preface to L.T.C. Rolt's (1961) account of the successful movement to save the railway, John Betjeman fulminates against the modernising trend that he associates with nationalisation, suggesting that the Talyllyn volunteers' success:

> is the result of the independent spirit which still survives in this country and refuses to be crushed by the money-worshippers, central-izers and unimaginative theorists who are doing their best to kill it.
>
> (Betjeman quoted in Rolt, 1961: xix)

Modernisation and rationalisation could at times be felt as a coldly remote and clinical exercise that took little or no notice of local needs or sentiment. Ironically it was the State that was criticised for its economic rationality, while the previous private *commercial* companies were remembered as benign custodians that had been untroubled by questions of efficiency.

In understanding the motivation for investment or any type of management decision on the railway, we must be aware of the complex and ultimately contradictory influences at work. Many of the lines and facilities on the network were duplicated, the result of competitive pressures on the industry in the nineteenth century, and were a historic legacy that the nationalised industry continued to bear. The Grouping of 1923 had effectively made these investments redundant. Perhaps the best example of this duplication was the Great Central Railway's London extension, opened in 1899. The line ran from Manchester via Sheffield, Nottingham, Leicester and Rugby, all of which already enjoyed access to the capital. Such a route could only be profitable by the abstraction of traffic from elsewhere (see Chapter 2). Although there were some initiatives to remove excess capacity, efforts in this direction were hampered by external pressures: from local MPs, from the Ministry of Transport, from local residents or customers and finally from within the industry both at shop-floor *and* management level. Thus, the financial state of many of the industry's services was fully appreciated some years before they were publicly exposed in the Beeching plan (see below). The strategy that BR adopted involved cross-subsidising loss-making services by maximising traffic on profitable ones. Put simply, this meant the maximising of gross income in order to maximise net profits. In many ways this was a tactic that railway managers had long deployed, and Hick (1991: 44) for instance notes the situation on the LNER in the interwar period:

> The measure of success or otherwise was, I believe, simply an estimation of receipts taken from the number of passengers carried by

each train. No real costings were then available – a full train was a good train.

The point was that in the absence of detailed statistics, it was assumed that an extra full or well-loaded passenger or freight train would add to the profit, or at least not make a loss for the organisation as a whole. The problem was that these extra or marginal services might well have been uneconomic when the full nature of their costs was exposed. The difficulty for railway management (that exists to this day) was that so many of the costs for one particular traffic were shared with others, and much of the capital deployed was sunk, so the 'true' cost was not amenable to simple calculation. It was this aspect of railway economics that was exposed in the critique offered by Beeching and others (Pollins, 1971; Joy, 1973; Hardy, 1989).

While the strategy of maximising gross revenue could work in an era of expanding traffic, its weakness was cruelly exposed with the rise of road-based competition and the growth in the level of rail-industry operating expenditure. As Gourvish (1986: 173) has highlighted, both sides of the railway's business were weakening in the 1950s. The share of total passenger mileage fell from 21 per cent in 1951–3 to 14 per cent in 1960–2, and on the freight side of the business, ton-mileage carried fell by 23 per cent, with market share falling from 45 per cent to 29 per cent. This decline, in spite of modernisation, led Government to look for new solutions to the industry's problems, including the introduction of non-railway management.

The Beeching era: 1961–5

The early 1960s saw yet more reorganisation of the transport sector in Britain. In 1962 the Transport Act was passed, which abolished the BTC and with it any pretence of the planned coordination of inland transport originally envisaged at nationalisation. The railway industry benefited from the creation of the British Railways Board (BRB), which had clearer aims and objectives. Growing losses by the industry and dissatisfaction with its organisational structure acted as a spur to change. To head the new Board the Minister of Transport, Ernest Marples, appointed Dr Richard Beeching, an industrial chemist with ICI. His only connection with the railways had been as a member of the Stedeford Committee created to examine the organisation of the industry in 1960, and he had sat on that committee precisely because of his *outside* business expertise (Bonavia, 1971: 101–27; Gourvish, 1986).

Marples' aims in appointing Beeching, and in the reorganisation more generally, were to set the industry clearer commercial directives by freeing it from historical obligations and restraints, and to bring in new managerial talent to the railways. The new chairman published his ideas on future strategy in the *Report on the Reshaping of British Railways* of March 1963, infamously referred to as the Beeching plan (Gourvish, 1986: pt. 3; Hardy, 1989). The report attempted to identify those areas that were loss-making and those that made a profit or which, with the necessary level of investment, could be made profitable. Beeching's aim was to rid the network of those traffics that it could not carry profitably, concentrating instead on fast passenger services and bulk high-speed freight, playing to the inherent strengths of the industry. The report identified those areas that were to be cut, including the widespread closure of many lines, services and stations. Here Hardy (1989: 73), a railway manager who wrote a sympathetic biography of Beeching, explains the cold logic behind the report:

> The studies clearly showed a remarkable variation in loading in dif-
> ferent parts of the country. One-third of the route mileage carried
> but 1% of the total passenger-ton miles of BR ... Similarly one-third
> of the route mileage carried 1% of the total freight-ton miles ...
> One-third of the stations produced less than 1% of the total passen-
> ger receipts and one-half produced a miserable 2%.

There was a parallel concentration of traffic on key routes and stations, with less than 1 per cent of the stations producing 26 per cent of passenger receipts, and the figures for the freight side replicated this pattern (Hardy, 1989: 73). Against this analytical, 'scientific' exposition of the facts, the previous management strategy looked amateurish, and the implication was that it had taken a non-railwayman, from the private sector, to identify and resolve the industry's problems.

The new chairman imported 40 officers from outside the industry between October 1961 and April 1963 (Gourvish, 1986: 337). In addition there were high-profile non-railway appointees at board level who were given functional responsibility. In many ways the influx of 'outsiders' as part of the Beeching era was but the latest instalment in an ongoing process that had begun with the nationalisation in 1948, but nevertheless there was considerable tension between 'insiders' and 'outsiders' during this period (Bonavia, 1981).

For the managers and staff outside headquarters in London there was a mixed reception for the plan, and again the understanding of the quality of this reaction is important in examining the manage-ment of the railways at this time. Beeching's cool impersonal logic meant redundancy for many railway workers. Staff numbers fell in the Beeching period from 508,000 to 380,000 (Hardy, 1989: 60). As Hardy shows this was appreciated by those asked to carry out the plan:

> But how did railwaymen feel? I was alright, I had a good job and, so far as I knew, good prospects, but what of those thousands of men whose lives were to be undercut by that long, long list of names of stations and lines to be closed? How could these men and women be expected to applaud the constructive elements? Many of them hated everything that Beeching seemed to represent.
>
> (Hardy, 1989: 69)

This tension illustrates C. Wright Mills' (1967) distinction between public issues and private troubles: there was a disjuncture between the formal rationality of a widespread closure programme for the future good of the industry, and the destruction of careers and lives that had been devoted to that very industry. The keenness of this loss is obvi-ously felt by Hardy, who had worked at the grass-roots level, but he could still see the validity of the aims of Beeching. On the shop floor, many railway workers suspected that traffic was being deliberately lost to the road-transport sector, but for management and supervisors too there was bemusement and frustration at having to turn business away and thereby lose revenue. There was great suspicion among staff as to the motives behind the Beeching plan, and these fears were heightened by the fact that Marples, the Minister of Transport who had appointed Beeching, had an interest in the road construction programme (see Henshaw, 1991).

Once again, the legitimacy of being (or not being) a railwayman had come to the fore. In many senses it was the fact that such identities were challenged or questioned that brought the issue into relief. The employment of many 'outsiders', coupled with the influx of several waves of consultants during the 1960s, made such identities discursive in much the same way that industrial action in the 1980s and 1990s was to do for the footplate and signalling grades. There are interesting parallels between the debates of the Beeching era, and more widely of the 1960s, and those of the 1980s, as to whether management rep-resented a generic practice or one that was dependent on the process in which it was embedded – in this case, the railways.

Coterminous with this attack on the established management of the industry was the self-conscious attempt to develop a new modern identity for the railways, a mood in step with Harold Wilson's modernising rhetorical zeal evident in his heralding the age of the white heat of technological revolution. Haresnape (1979) describes the way BR tried to refocus its image during the late 1950s and 1960s, in part because of the growth in competition from road and air services. Symbolically, the most important moment of this process was the re-equipping of the west coast main line from Euston to north-west England and Glasgow. Much was made of clean electric traction used on the route and the rebuilding of the London terminal, with the associated destruction of its famous Doric arch, now viewed as the touchstone of postwar architectural philistinism (see Pearson, 1967: chs 13 and 14; Dixon and Muthesius, 1985: ch. 3; Richards and MacKenzie, 1988). There are interesting aesthetic disjunctures between this modernising impulse and later image-building within the privatised industry, which looks to much earlier periods for its inspiration. Samuel (1994: 77) has drawn attention to the irony:

> Where the 1950s and 1960s were good at making the old look new, the 1970s and 1980s were no less resourceful at establishing what was called ...'instant oldness'.

Paralleling this modernisation was the development of the railway preservationist movement and an allied heritage publishing industry that nostalgically celebrated this particular aspect of Britain's industrial past. The rediscovery of this heritage was symbiotically related to the modernisation process itself (see Kellett, 1969; Whittaker, 1995; Wilson, 1996; Payton, 1997; Sykes *et al.*, 1997). John Betjeman (1972: 8) wrote in a book celebrating railway architecture:

> Already, I suspect, both our books are mainly of historic interest, for the architects of British Rail never cease to destroy their heritage of stone, brick, cast iron and wood, and replace it with windy wastes of concrete ... I became aware that the rich inheritance of railway architecture from Victorian pride in achievement to Edwardian flamboyance, declining to the poverty of the new Euston, has a moral. There was nothing modern to photograph.

Organisationally this mood of modernisation, with all its contradictions, is captured in Fiennes' (1973) autobiography. He discusses the aftermath of the shake-out of the 'Old Guard' when he took over as general manager at Paddington from Stanley Raymond, later to become chairman of BRB:

> He [*Raymond*] had gone like a destroying wind through the traditional practices of the Great Western. He had symbolically stripped the works of Brunel and Pole and Milne from the Board Room and the corridors down to the basements and cast the attitudes to the four winds. The staff were at odds about it all. Many were frankly bolshie; many were dazed; some were frightened; some were in transition from those three states of mind to the knowledge that something very useful had happened to them; very few were wholeheartedly on Raymond's side.
>
> (Fiennes, 1973: 94)

Fiennes, like Hardy, was a manager who was sympathetic both to modernisers and traditionalists in the debate over the rationalisation of the industry. Both saw the necessity of changing patterns of service, but both had been socialised as managers into a railway culture that valued 'being a railwayman'. Both were clearly in an ambiguous position as they saw the logic of the new order but also realised they were embedded in a system of management governed by norms of reciprocity. There was resentment too at the implication, clearly evident in many of the autobiographies, that the organisation and its antecedents were not business-like, or that they were doggedly clinging to tradition. Many managers and officers had worked for the prewar commercial companies, in which attempts to modernise had been compromised by lack of capital rather than inertia.

The tension between 'insiders' and 'outsiders' can be examined in terms of notions of emotion and nostalgia within the organisation. Fineman (1993) has stressed the importance of studying the organisation as an emotional arena rather than seeing such subjective factors as simple variables. As Putnam and Mumby (1993: 36) have argued:

> Emotion, then, is not simply an adjunct to work; rather, it is the process through which members constitute their work environment by negotiating a shared reality.

Some of those who considered themselves career 'railwaymen' within BR management were emotionally attached to their occupation in a way that those coming new to the industry could not be.

Gabriel (1993) has argued that nostalgia acts as a resource that older or marginalised members of an organisation draw on when their identity is challenged, and such emotion therefore acts to legitimate pre-existing knowledge and cultural practice. The opposite of nostalgia can equally be used as a resource. Gabriel employs the term 'nostophobia' to chart the way some organisational actors demonise, rather than celebrate, the past. The workplace becomes a battleground where competing collective versions of the past are traded. While such explanations add a richer understanding of individual and group behaviour within the organisation, it is important to stress that organisations are more than the sum of opposing discourses. These discourses are themselves materially linked to practices in the organisation over time, thus our use of such conceptualisations once again has to be embedded within the complexity of historically situated social forms and actions.

Bonavia, himself a career railwayman, eloquently discusses the tension between 'insiders' and 'outsiders':

> There was however one consequence of the new style of 'business management' for which the 'Beeching boys' were largely responsible. This was a reaction against any concept of public service and even more against the idea that anyone could be in railways because they liked railways. That was 'playing trains' and it was supposed to have been a fault of past generations of railwaymen.
>
> (Bonavia, 1981: 133–4)

Therefore the very love for the job, the embeddedness of the management individually and collectively in the work, was seen as a disadvantage and, indeed, the very cause of the parlous state of the industry. Such criticisms of the nature of railway management fundamentally challenged the basis for identity of those who aspired to be 'railwaymen'. In many ways the challenges to occupational identity in this period foreshadow the wider attack on public-sector management that began in the late 1970s, when to be a worker or in management in the public sector was viewed as proof that the individual had failed to find a job in the private sector – it was not viewed as an active career choice (Pollitt, 1993).

It would be wrong, however, to imagine that the influx of new managers, consultants and other non-railwaymen destroyed the established 'culture of the railroad', as Gourvish (1986: 388) has labelled it.

Bonavia (1981: 133) describes the grudging respect that those 'inside' latterly received from those 'outside':

> the 'old hands' decided that the quality of the new blood was variable. It would be an over-simplification to say that the new men came to scoff and remained to pray; but if they had expected to find railway managers fumbling with problems that only needed firm handling, they soon learnt that this was not so, and that the general level of competence among railwaymen was at least as high as in large-scale industry generally.

Bonavia touches on the same issue later, and notes the transitory nature of the newcomers' presence:

> Tensions sometimes were felt between the Beeching 'new men' and the 'regulars' of the railways. But in many cases a modus vivendi was established, and some of the 'hard-faced men' from industry identified themselves closely with their railway colleagues. In a few cases, a sigh of relief went up when the stranger within the gate decided to return to the outside world.
>
> (Bonavia 1981: 218)

The resentment felt about outside interference in the industry was very real and is exemplified in a reported conversation that Peter Parker had with the then vice-chair of BR, Bill Johnson. Parker was sounding out Johnson after being offered the chairmanship of BR by Minister of Transport Barbara Castle in the late 1960s, a position that Johnson was subsequently offered and accepted:

> Johnson, at sixty, made no bones about his opposition to an outside appointment. Railways needed a rest from political interference: that was the attitude of the railway managers. They were fed up with outsiders coming in, getting a gong and getting out. No railwayman could get to the top through the scramble of politicians, civil servants, consultants – and 'outsiders'. I interrupted, genuinely puzzled, 'What about Sir Stanley Raymond?'
>
> 'He's not a real railwayman, he's only been with us for ten years, he came from the buses.'
>
> (Parker, 1991: 146)

After Beeching

Beeching left the industry in 1965, a year after the re-election of the Wilson administration, and was replaced by Stanley Raymond (Gourvish, 1986: 344–74; Hardy, 1989). This did not, however, usher in a period of stability following the turmoil of the Beeching years. The closure programme continued, albeit at a reduced rate, and with it many thousands of railway workers were made redundant (Bagwell, 1982; see also Wedderburn, 1965). New priorities were set by the Labour Government for transport policy, culminating in the 1968 Transport Act (Bonavia, 1971: 128–32; Bagwell, 1982: ch. 5). This legislation finally gave recognition to the public-service aspect of the BRB's losses by allocating a grant for running socially necessary but loss-making services. The Act also wrote off a large amount of debt incurred by the Board, and finally made a distinction between profitable and loss-making sections of the freight side of the business, with the latter being transferred to a separate body. The aims were to set the industry on a new footing with profit and loss clearly separable, and to encourage management to act commercially. In spite of these reforms, and those of Beeching earlier, the financial position of the railways continued to deteriorate. True, costs were falling and productivity was rising, but revenue was declining at a faster rate. This was due to a combination of factors: a switching of freight to the roads, a reduction in some flows and the ongoing rise in private motoring, itself part of the long postwar boom with its associated affluence (see Dyos and Aldcroft, 1974).

In addition to the major reorganisations mentioned above, the BRB also brought in a series of management consultancies, which were to advise on the possible shape that the management and board should take. Gourvish (1986) tells of the damning report on regional management in the North Eastern Region in the late 1950s, prepared by Urwick, Orr and Partners. At national level the influence of outside consultancies was felt, with Cooper Brothers, Production-Engineering (in 1966) and McKinsey (during 1968) all carrying out studies. It is interesting to note that Cooper Brothers, later Coopers and Lybrand, would become one of the main consultancy firms in the privatisation process of the 1990s (see Chapters 5 and 7). Major enquiries into the nature of the organisation were held during this period: in 1960, 1967–8, 1973 and 1976–7. At times these reorganisations simply reinvented what had previously existed. Some of the efforts of Beeching to develop functional responsibility, for example, had *recreated* a strong departmental form of control that had previously been viewed as mistaken (Gourvish, 1986: 340–1). On the ground, management considered this change at best as an unwelcome distraction. Fiennes (1973: 74) juxtaposes his pre- and postnationalisation experiences:

> I reflect now that pre-war the L.N.E.R. set up an organisation in 1923. It stood the test of time, give or take a little, until 1947. Since then I have been re-organised in 1948–9, 1953, 1955, 1956–7, 1960, 1962, 1964, 1966–7.

Fiennes also gives an indication of the bureaucracy associated with the position of a chief operating officer in the 1960s organisation, to which the kinds of reorganisation alluded to above only added:

> After a fortnight, Ena [Fiennes' secretary] and I added up the score. The average number of sheets of paper through her to me each day was 292. The number of days on which I should be sitting on a committee was 212 out of the next 250.
>
> (Fiennes, 1973: 75)

Mapped onto this uncertainty (in terms of organisation at the macro and mezzo level) was political instability: there were 11 different Ministers of Transport between 1948 and 1974 (Gourvish, 1986). This post, although sometimes part of the cabinet, has traditionally been seen as a junior one, used as a stepping-stone to more exalted briefs – a trend that continues to this day (Bagwell, 1996). There was also a tendency by all the postwar Governments to use the nationalised industries as instruments to control the economy, either by increasing, or more usually, by cutting investment levels. Such moves were often made at very short notice and played havoc with investment plans that might take nearly a decade to come on stream. The railways were also used as a tool in the fight against inflation, with Government interfering in the prices charged for both freight and passengers (Joy, 1973; Gourvish, 1986: 185–6).

Despite all of these organisational changes, there remained a strong identification with the industry from within the management and supervisory grades. Many senior officers and lower level management were career railwaymen who had worked in the industry for much of their adult life. At the end of the 1970s, the chairmanship was given to Peter Parker, who had spent his career in private industry. His autobiography gives a fascinating account of this railway culture at all levels of the organisation, as well as his own immersion in it. Several years into his appointment, Parker felt he had become part of the industry in much the same way as Elliot before him had:

> I had become a passionate railwayman, with the special zeal of a true convert. I was enjoying the years of comradeship in BR. I had never felt alone. 'The railway community' was a vision that I found compelling ... I believed in what I was doing and I loved it.
>
> (Parker, 1991: 251)

The problem for the industry, and those who worked in it, was that as the 1980s loomed the railways were faced with difficult options and choices. Many of the productivity gains that had been realised over the previous 30 years had been made in an environment of full male employment and at a time of economic growth. The 1970s, however, witnessed a challenge to this stability. There was a combination of a downward pressure on wages in the industry as inflation began to bite, along with a further loss of traffic. The end of the 1970s saw the election of a Conservative administration whose leader and close advisors were actively hostile to the public sector as a whole, and the railway industry in particular. This combination of events coincided with the need to renew much of the investment that had been made in the 1950s. All these factors meant significant restructuring in the industry was inevitable. Before moving on to examine the period after 1979 in greater detail, we need to understand the experience of the workforce during the first three decades of nationalisation.

4

A Seat at the Table? Working for the Nationalised Railway, 1948–79

> [T]he main problem was recruitment to fill vacancies – recruitment, that is, of suitable staff who would develop into railwaymen of calibre equal to those of the pre-war period
> (Michael Bonavia, *The Birth of British Rail*, 1979: 86)

Introduction

On my study wall there is a black and white photograph of six sooty men on the buffer beam of a locomotive (see Figure 4.1). Each of them has his grease top cap (the badge of the locomotive department) raised high in his hand, and a wide grin etched on his face. The caption reads: 'Footplatemen celebrate nationalisation at Newcastle Central station, 1st January 1948'.

At the start of 1948 over six hundred thousand employees of the four former private companies became workers in the State sector. This period was marked by both its stability and simultaneously by change. Railway workers whose career spanned this era, including those at Newcastle that day, witnessed the growing pains of State ownership, the upheaval of modernisation with the introduction of new technology, and the rationalisation of the network. In addition, throughout the period there were subtle shifts in the culture of work and in particular the identity and status of railway employment. In this chapter I explore all these separate but ultimately related issues through the accounts of those working in the industry, of academics and of other commentators. The aim here is to understand how change was experienced both on the shop floor and at a wider cultural level. The chapter builds on earlier discussions of the way occupational identity in the industry was formed and reproduced over time and develops an

Figure 4.1 Footplatemen celebrate nationalisation at Newcastle Central Station, 1st January 1948. (Ken Hoole Collection, Darlington Railway Museum)

understanding of railway culture at the cusp of the Thatcher era, which is discussed in Chapters 5 and 6. It is argued here that many of the later reforms of the 1980s and 1990s were based on a particular, and rather crude, reading of labour and industry in general during the postwar period, and that railways in particular provide an ideal way in which to view this process in action. The chapter is divided into five sections, which look at:

1. the early experience of State control
2. modernisation and the loss of skill
3. the effect of closure and in particular the Beeching era
4. the sense of decline in railway labour
5. the issue of lost status.

Nationalisation

While my photograph from Newcastle may have been posed for the press, the workforce and their trade unions, as Bonavia (1981: 20) notes, met the advent of State ownership with general acclaim:

> So far as the rank and file were concerned, while there seemed to be no immediate change in their work, those who were keen union members were pleased that nationalisation marked the conclusion of a campaign waged for many years by the railway trade unions. On the Sunday following the changeover, the National Union of Railwaymen held mass meetings in many large railway centres. More than 2,000 railwaymen attended the London meeting in a West End theatre, addressed by the Union's President.

Bonavia goes on to list a number of events held to mark the commencement of State ownership, including 'a smoking concert' held by Stourbridge No. 1 Branch. Norman McKillop, author of the official history of ASLEF published in 1950, wrote in glowing terms of the transformation that nationalisation heralded:

> Out of a chaos of almost accidental undertakings the whole of the railway resources were now directed to one end. The ideal we had pursued down the years was now an accomplished fact. We also had achieved what our founders had set out to do ... The 'old unhappy days' are gone forever, and in their going have left us also with a changed approach. Even before this great event of nationalisation

we had developed a new sense of responsibility, a new assessment of those who sat at the other side of the negotiating table; by 1948 we had long known that the problems were not confined to our side alone, and this new set-up would merely mean a closer understanding, a new feeling – that we were at last sitting 'round the table' and not on the opposite side to that of our new employers.

(McKillop, 1950: 359–60)

McKillop's prose is heavy with expectation about nationalisation. State ownership is interpreted as the unfolding of an inevitable rational process, the destination of the 'forward march of labour'. Organised labour had won its seat at the table and was now an equal partner in building the future. However, as we saw in Chapter 3, a detailed discussion of what nationalisation actually meant in practice was seen as unimportant. As Tomlinson (1982, 1997) has argued, the problem of what form State control would take was resolved by the adoption of the Morrisonian public corporation model – as in the case of the London Passenger Transport Board (LPTB) – and the rejection of any meaningful form of workers' control or involvement. While there remained those in the labour movement who were critical of this enthusiasm for technocratic solutions, official union policy was to back the Government's position. As Howell (1999: 346) says:

many trade union members viewed the 1945 Government as 'their' Government in a way that was unique ... strongly positive emotions [towards Labour] could bolster trade union support for potentially unpopular policies.

While there was not the same level of hostility and bitterness towards the 'old companies' as had been the case in the coal industry (see Powell, 1993; Outram, 1997), it would be wrong to suggest that there was no ill-will felt between railway workers and their former employers. Harry Friend, a fireman and driver in various sheds in the northeast who began his career with the LNER in the 1930s, spoke of the legacy of bitterness that flowed from the private companies' response to those of their workers who had taken industrial action during the General Strike in 1926:

We had had a lot to put up with the private companies. When they had the General Strike in 1926 in the eastern region, when the drivers and firemen came back to work they had to work a three-day week and they lost three days. It was a six-day week then, not a five-day week ... and the cleaners at Durham were put on to a two-day week and they were married men. So basically when we were nationalised we thought it was for the better, we were rid of the LNER and it's now the British Transport Commission. So it was an improvement.

(Harry Friend, interviewed 1995)

But in spite of the transition into public ownership, little had changed for the workers, echoing the experience of lower levels of management seen in Chapter 3. As Harry Friend noted later in the same interview:

One day we worked for the LNER, the next day it was BR. The wages weren't altered or anything like that. We thought naturally working for a Government concern ... it was a better thing.

These twin themes of expectation and little actual change are repeated in several of the interviews I carried out. Tom (Tiny) Clementson, a fireman at Heaton shed in 1948, discusses his memories of the period:

Well it didn't make any difference to our department. Them that had worked for the old company before the war thought national-isation was going to make a big difference, but it made very little difference to us. You used to hear tales of before the war when the depression was on and the chief clerk used to come down (they used to call him the good news clerk!) come down and give them their notice like, they were made redundant, finished.

(Tom Clementson, interviewed 1995)

Ron Bradshaw, a signalman on BR London Midland Region (LMR) and later the Western Region (WR), compares the optimism felt by some with the pessimism of others:

Young socialists like myself now eagerly and confidently awaited announcements of wage increases and conditions equivalent to postal workers and other civil servants, the promise of enterprising plans for modernisation and expansion, and electrification of all main lines ... The more mature of our compatriots were far more sceptical.

(Bradshaw, 1993: 106)

These different attitudes towards, and experiences of, nationalisation have to be put in context within the historical moment and life cycle of the staff involved. For those who had worked for the former private companies before the war there was a sense in which that period was one marked by depression and retrenchment. The industry had seen lay-offs and redundancies as well as industrial bitterness in the wake of the General Strike (see McKillop, 1950; Bagwell, 1963; McKenna, 1980; Potts, 1996). However, for those workers who had entered railway service just prior to or during the war, the experience of work had been qualitatively different. One possible explanation for this changed experience is the increase in the strategic power of labour as a consequence of the beginning of World War II, and the associated tightening of the labour market. Most of the older drivers I interviewed had entered railway service in the period 1937–42. Such was the demand for labour in the locomotive sheds of the north-east that many of the interviewees said they could never remember cleaning engines at all – cleaning, as we saw in Chapter 2, being the most junior rank in the footplate line of promotion. Instead, they were quickly given the status of passed cleaner and were employed as firemen on a semipermanent basis. This was in contrast to the pre-war years, when a new entrant to the same department might have to wait a decade to be a fireman (McKenna, 1980; Farrington, 1984). The war, therefore, seems to have represented a break with the experience of work during the 1920s and 1930s, and in some senses this ameliorated the collective bitterness towards the old companies. As we saw in the previous chapter, managerial attitudes and practices towards blue-collar grades changed under the stress of war and this continued into the postwar period (Fiennes, 1973; Hardy, 1989). The frontier of control between management and labour had never been simply the cash nexus; rather it was always contextual, overlaid with a range of forms of rationality. The experience and demands of wartime made this relationship still more complex.

Several of my interviewees noted the way in which the original optimism about State ownership was quickly dashed. Gilly Young had started his railway career at Rothbury shed in Northumberland, but was at Heaton by the time of nationalisation:

> We all thought at the time it was the saviour of Britain's railways, of the railway system as we knew it ... thought it could do nothing but good; it was a good idea at the time but it wasn't allowed to work out that way. But slowly when the Tories took over it was all

chopped away you see, so it was a good idea in the early days, but it was destroyed.

(Gilly Young, interviewed 1995)

Other interviewees spoke of the way the levels of bureaucracy quickly increased after nationalisation. George Deownly was a fireman at Botanic Gardens shed in Hull at the time:

> It [*nationalisation*] was a good thing, the only thing was they made too many chiefs. Instead of sticking to the same kinds of staffing arrangements that was already there, they increased the office side. And as the staffing arrangements went up, we only had a certain amount of inspectors and the inspectors increased as well.

(George Deownly, interviewed 1995)

This latter point is interesting in a number of ways. Firstly, it highlights the impact of the increasing bureaucracy at the micro level as a result of the corporate form adopted with public ownership in 1948, something which is reflected in the historical record of the time (Bonavia, 1971; Gourvish, 1986). Secondly, it is indicative of a class tension played out on the terrain of the 'productive' and the 'nonproductive'. The office workers are by implication white-collar workers and therefore something 'other' than the manual working-class railway operating grades who see their work as the productive, necessary 'actual' work. Penn (1986), in his discussion of the socialisation of workers into skilled identities, talks of the way such groups set themselves apart from white-collar or managerial groups, as well as from the unskilled and other trades (see also Eldridge, 1968: ch. 3; Steiger, 1993). It could also be argued that this interpretation is echoed in Goodrich's (1975) analysis of workplace custom and tradition. Goodrich talks of the limited horizons of British craftsmen, who wanted to be left alone rather than actively engage in workers' control, but 'by the very pride in that individual skill it stiffens the refusal to be controlled' (1975: 41). The implication from the above is that had the workers been allowed greater autonomy, nationalisation could have been a success. Or, to put it another way, it was the poor level of management in the industry that created many of the later problems.

State ownership, therefore, initially changed very little for staff in the industry on a day-to-day level. The detailed labour process remained much as it had during World War II, as traffic levels remained high and little capital investment was being made. But this was to change as British Railways finally began to modernise as a result of a huge injection of State finance in the mid-1950s.

Modernisation and deskilling?

In 1955 the British Transport Commission (BTC) published its Modernisation Plan, which aimed at the complete eradication of steam traction, still the dominant form of motive power in the industry at that time, within 15 years (Haresnape, 1981; Gourvish, 1986: ch. 8). This investment, alongside the rationalisation of the system (see 'The Beeching era and closure' below), had profound implications for the workforce. At the start of this period there were over six hundred thousand employees in the industry, while by the late 1970s this figure had fallen to 182,031 (Bagwell, 1982). The premodernised industry was highly labour intensive: in 1950, for example, there were over ninety thousand workers employed in the footplate grades alone, 18,000 of whom were men and boys involved simply in the day-to-day maintenance and repair of steam traction. Bagwell (1982: 52) gives an indication of the range of occupations associated with steam locomotives:

> There were engine-cleaners and chargemen engine-cleaners; boiler-washers and chargemen boiler-washers; coalmen; firedroppers; steamraisers; locomotive shunters; cranemen; shedmen; store-keepers; timekeepers; hydraulic and pumping-engine staff and the water-softening plant attendants.

By 1978, the number of drivers had fallen to 26,000 (Bagwell, 1982: 45).

Similarly, in 1950 there were 25,190 signal workers employed on British Railways. By 1978 this figure had dropped to 7,961, a fall of nearly 70 per cent (Bagwell, 1982: 44). Like the locomotive department, in 1948 the signalling system was heavily dependent on labour and mechanical methods of operation. The fall in staff was as the result of rationalisation and modernisation, with modern power signalling boxes capable of controlling far wider areas with far fewer staff.[1]

Although objectively modernisation was to deskill the collective workforce, this was not uppermost in the minds of those involved in implementing the plan. Indeed, the introduction of diesel traction was far from unproblematic in terms of skill reduction. There were some grades that suffered 'death by dieselization' in Cottrell's (1951) phrase. But other workers saw their skill levels increasing as motive power became reliant on electromechanical and subsequent electronic

Figure 4.2   *On Early Shift* – Greenwood Signal Box, New Barnet. Promotional poster for BR by Terence Cuneo, 1948. (Cuneo Fine Arts/NRM/Science & Society Picture Library)

equipment. This contradictory tendency of simultaneous deskilling and upskilling is noted in Penn's (1994) work on the formation of skill in the UK labour market during the 1980s.

The drivers involved in the modernisation process were trained on the new traction by instructors employed by British Railways rather than through the Mutual Improvement Class movement (MICs), as had hitherto been the case (see Chapter 2). Harry Friend (1994b: 242) recounts the initial training course in the north-east area:

> The principal looked after the class room instruction. Five days were allocated for this activity out of a fifteen-day course. The remaining ten days were spent 'outside' on a static Class 40 [*one of the Modernisation Plan prototype diesel locomotives*] for three days before going into traffic for handling experience purposes.

Friend goes on to highlight the implications for the drivers involved:

> A class consisted normally of eight drivers and two practical in-structors. In retrospect it was a turbulent period. The drivers in the beginning were the senior hands at Gateshead depot, men in their sixties. To sit for five days in a classroom and be confronted with basic electricity ... cumulators, heat compression engines and so on proved to be wildly traumatic in many cases. Some were totally bewildered and never came to fully understand diesel-electric traction.

The novelty of the situation in which many drivers found them-selves was reflected during an interview with Bob Howe, a former footplateman, who at the time of dieselisation was based at Tyne Yard:

> It was all strange. We were steam men, and they suddenly chucked a diesel at you like, and you had to learn it. I found when you eventually got to know them and work with them they were better to work than steam engines; they were easier and more comfortable.
>
> (Bob Howe, interviewed 1994)

Within the formalised training scheme there was still a major role for the kinds of informal knowledge and tacit skill that had been a prominent feature of steam traction. The drivers who were converted to diesels still had to make use of this informal knowledge if they were to do their jobs effectively:

> Really the company got away with murder, learning us the diesels. When they introduced the diesels they just put us through as quick as possible – 'yer passed'. Well really when we were passed there were times when we were lost. Fault charts [officially provided manuals] were useless. You found out eventually through experience, sitting and listening in the messroom. 'Oh I failed', and someone would say 'Well what did you do?' ... 'Oh I did such and such', and you kept that in your head, and sure enough it happened.
>
> (Bob Howe, interviewed 1994)

And a colleague of Bob Howe who had experience of the process of dieselisation at Tyne Yard took up this theme in the same interview:

> You always remembered that [*informal, tacit advice*], you remembered that more. I mean you got all the fault charts out, trying to fathom the fault out on the fault charts – would take you half an hour to fathom the charts out, they were that complicated. Since then they have simplified them, but only through the likes of Bob's era getting into trouble.
>
> (Morris Mowbray, interviewed 1994)

This account, and others obtained through interviews, emphasises the importance of the oral tradition coupled with personalised knowledge in this sort of work. The worker proffering the advice is seen as legitimate by his work mates. Despite the fact that formal training was organised by the employers rather than by the autonomous, craft-based MICs, the notion of self-help within the grade remained a central and vital aspect of work culture, and was crucial to the successful introduction of the new technology. Morris Mowbray gives a vivid description of the unofficial fault-finding and problem-solving with regard to the brake system adopted by some at Tyne Yard:

> You would be sitting in the middle of the mess room and somebody would say 'Hey I shoved a pencil up the middle of the RCR and we got vacuum straight away and away we went'. But you always remember that. And then someone would say 'Hey he put a pencil up the bugger the other day I'll gan and try it'. But all these things

was through fellas getting into trouble, through failures, built the fault charts up that way. We were thrown in the deep end really.

(Morris Mowbray, interviewed 1994)

The footplate grade, therefore, built up incremental knowledge of this new technology through active involvement in the labour process. In this sense modernisation did not represent a neat break with the past. Rather it was precisely because of the continued existence of a strong autonomous workplace culture within the footplate and other grades that change occurred as smoothly as it did. Thus in no sense could the movement from one form of traction to another be interpreted as a crude attempt by management to deskill the workforce.

The complexity of the issues facing management besides the direct control of labour has been recognised by others. Tomlinson (1982) is critical of those engaged in the labour process debate for privileging this aspect of management's role above any other. In doing so he believes that theorists have produced a functional account of management, which ignores:

> Problems which govern the activities of most managers – marketing, cash flow, supply of components, quality control, etc. – [which] are striking by their absence because they are not readily assimilable to the assumed over-arching question of the management of labour.
>
> (Tomlinson, 1982: 25–6)

The apparent collective reduction in skill that dieselisation represented for the footplate grade must be placed within the context of complex and often contradictory management aims and goals. The pressure on BR management to realise productivity gains has to be read against the factors of cheap fuel oil in comparison with coal, the tightening labour market in key parts of the country, and political and patriotic pressures to source the majority of the new forms of traction from inside the UK. This last point resulted in the ordering of numerous untried designs from a wide variety of domestic manufacturers who had had little or no experience of new technology. Regional autonomy within British Railways itself added to this complexity. The Western Region, for example, chose to order nonstandard types of locomotive. This lack of standardisation was recognised by footplate staff at the time:

> The steam engines basically were all the same, you could get on any steam engine and it was like second nature, you could drive it

straight away ... But diesels were slightly different in regard to failure, switches and stuff like that, relays, release buttons for the brakes and that, all in different places. They could have standardised them more. Type 4s and the type 3s both English Electric, both nose ends, so the same switches could have been put in them, but they weren't, they were totally different ... you know it just made it more complicated for you to remember.

(Morris Mowbray, interviewed 1994)

The footplate grades therefore continued to enjoy relatively high levels of skill and autonomy in the wake of dieselisation. In addition to the changing technical competency referred to above, there remained other elements of skill such as route knowledge that meant drivers retained considerable strategic power over the labour process. This status was further buttressed and institutionalised through the footplate unions and the highly formalised industrial relations system. Control over work was exercised by local union representatives in negotiation with managers through the planning of rosters and the allocation of duties at and between depots.

The introduction of new technology was less straightforward for the signalling grades, and experienced very unevenly. Some found their job changed very little, with older mechanical signalling remaining in day-to-day service throughout the period. But others experienced the abolition of vast swathes of individual boxes, that were replaced by centralised control in a modern power box, with one new centre controlling the area of perhaps thirty or more of the older installations. But when modernisation did occur, the skills and knowledge of signal workers were not necessarily lost. The concentration of control into larger power boxes demanded greater levels of knowledge on the part of the individual worker, because of the larger area and the increased number of trains and locations supervised at any one time. As Pendleton (1984: 35) has said:

The organisation of work arising from technical change in signalling, however, does not necessarily bear simple comparison with manual operation, and whilst some aspects of decision-making may be lost by signalmen new forms may be gained. Signalmen's role in the conception and execution of tasks is thus best seen as changed rather than simply increased or reduced. Increased control of the 'production' process as a whole, then, does not have any necessary implications for labour control.

We see, therefore, that the experience of technological change that occurred under the Modernisation Plan was far from straightforward. The transition was not a crude attempt on the part of management to enact deskilling in order to gain greater control at the point of production. The important point to make here is that the transition to new technologies was rendered intelligible precisely because there was a strong workplace culture which shared and valued a collective knowledge of railway operating procedures and practice. But this change in technology did not occur in isolation, and it is to the issue of rationalisation that we now turn.

The Beeching era and closure

The period of Richard Beeching's chairmanship of BRB (1961–5) and the implementation of his now infamous plan represented an important watershed for railway workers in general, as well as for the footplate and signalling staff discussed above. Insecurity was felt by many for the first time in their careers as a result of both the Modernisation Plan and the closure programme that was accelerated at this time. For workers in the industry, the impact of rationalisation was experienced differently in separate parts of the country, as BR gradually withdrew from particular types of traffic or closed whole lines (Bonavia, 1971, 1981; Gourvish, 1986; Hardy, 1989). Stan Fairless, a signalman in the north-east at the time, remembers having to fill in the traffic survey on which Beeching's 1963 *Report on the Reshaping of British Railways* was to be based:

> I was at Teesside at the time and each signal box received a graph to fill in of traffic positions over 24 hours ... all these forms were sent to a central point ... so certain lines that fell below a point on the graph, they closed the line, didn't matter about connections, or feeder trains. Beeching was a hatchet man, any silly bugger could have done what he did.
>
> (Stan Fairless, interviewed 1995)

This sentiment was repeated by many of the retired railway workers I interviewed. George Deownly, a passed fireman at the time of Beeching, discusses the closures and their effect:

> Anybody could have done what Beeching did. That man was not a railwayman for a start. Railways were an asset to the public, and

branch lines like Hornsea, Withernsea, Alston, lines I know, they closed them and when they closed these branches they didn't think of the people, the people did not exist. They thought that everybody had a motor car.

(George Deownly, interviewed 1994)

Walter Mulligan, another driver at the time, also discussed the human effect of closure, and his account is interesting in the way organisational change is remembered through the personal:

It was a scandal: close this line, close that line. And there was men just literally thrown on the streets. Signalmen, they lived in the area and they were brought up in the area, they were local people, and the little local sheds were all closed eventually.

(Walter Mulligan, interviewed 1994)

This stress on the individual and the community effects of closure and rationalisation are important and are reflected not only in interviews carried out but also in the numerous autobiographies of railway workers who lived through the period (see for example Smith, 1972; Vaughan, 1984; Newbould, 1985). Emphasis is placed on the community and the stability that the railway represented. In many ways the rural railway with its country station represented the community's link with the modern world. The brash newcomer of the mid-nineteenth century had become the slightly decayed but deeply embedded fixture of many settlements, and its workers important members of the community. Ironically, the steam railway in the countryside becomes part of an iconic vision of timelessness, the invocation of Englishness (see below, also Payton, 1997; Sykes *et al.*, 1997).

The best illustration of this sense of embedded historical continuity is found in the series of autobiographies by Adrian Vaughan, who worked on the Western Region (WR) of BR from the 1950s (Vaughan, 1984, 1987). His writing, tinged with a romantic nostalgia, reflects the continuity of railway employment across the years, describing the kind of labour he carried out in his early career as virtually unchanged from that of his predecessors at the turn of the twentieth century. Vaughan worked alongside several workers from much older generations, his socialisation taking the form outlined in Chapter 2. His readers enjoy a fascinating portrayal of 'work as craft', and the sense in which there is a collective moral responsibility for occupation passed from generation to generation. What is so telling here is that Vaughan desires to replicate in his own career the lives of his heroes, and yet knows that this world is increasingly under threat (Vaughan, 1984: chs 4–6). Juxtaposed with this Elysian *fin-de-siècle* vision of rural bliss, Vaughan describes his 'perfect job' being destroyed before his very eyes by a mixture of modern technology and nonrailway management closely associated with the Beeching reforms. In his chapter concerned with this period, fascinatingly entitled 'Progress the ogress', Vaughan compares his traditional unchanging railway with the new:

> Steam engines and semaphore signalling were the heart and soul of the Western Region because the two systems generated the morale in the men that ran the railway ... As a result the men's morale was high as they worked in the satisfying knowledge that they had a difficult job under control. The railway was a team of long-standing friends, the work was permanent and there was a feeling of good fellowship and security ... Then came modernisation ... I watched events from my signal box and felt the fires of outraged loyalty flame evermore furiously in my coal-burning heart as my railway was slowly dismantled – at great expense – in what seemed to me to be a quite unnecessary search for efficiency. We already had it.
>
> (Vaughan, 1984: 328–9)

There are strong parallels between Vaughan's antimodernist writing and those of Tom Rolt (1978), who wrote extensively on industrial archaeology during this period. Ironically, both were fascinated with the products of industrialisation, but were profoundly disturbed by its contemporary developments. In both writers' work it is not technology per se that is objected to, but rather its introduction at the behest of remote bureaucrats unaware of either the technology or the social relations that were being threatened by change. This detachment is illustrated in the comparison Vaughan makes between the management qualities of the two eras:

> In Hank's [*former Chairman of WR*] day Western Region ran a complicated network at or nearly at a profit, officials were very busy and those few who visited outlying stations such as Challow did so by train but in the years of decline after 1960 there appeared more frequently what I took to be the 'New Men' of the railway – and they rode in chauffeur-driven Bentleys.
>
> (Vaughan, 1984: 343)

Figure 4.3   Clear Road Ahead. Promotional poster for BR by Terence Cuneo, 1949. (Cuneo Fine Arts/NRM/Science & Society Picture Library)

Vaughan's bitterness over this change was based on his view that the railway was deliberately being run down in order to make closure of lines or services easier. He saw this as being due to the new type of managers who were employed at the time, with no interest in the job in a vocational sense, and who were simply unaware of the 'errors' they were making:

> The ranks of the new supervisory grades were being filled by recruits from universities and by the more ambitious booking clerks. After a two-year course these men – or boys – could take charge of railway-men who had been in the service for forty years or more.
>
> (Vaughan, 1984: 344)

There is here a clear moral outrage at the way less experienced managers seek to downplay the importance of seniority and knowledge of older railway workers. At one point Vaughan is told by a young manager, to whom he was extolling the virtues of the 'old hands': 'You go a lot on these old blokes, but really their job was no more than one year's learning and forty-four years' repetition' (Vaughan, 1984: 344).

Many of my interviewees and those writing their reminiscences dwell on the subsequent loss of traffic during the 1960s and into the following decade:

> Returning to Summerseat box for the second time, I took as a matter of course. Seven years had elapsed and the drop in traffic flow came as a shock. Freight had been reduced by one-half and passenger services had been cut back considerably.
>
> (Bradshaw, 1993: 113)

Decline in the industry was thus felt in a very personal way, and was linked to the immediate events and the characters of those introducing the changes. The new type of management, whose presence was explored in the previous chapter, was held to be responsible for much of the loss in traffic and services. Importantly, the 'new men' were blamed because they were not 'of' the industry, they were not 'railwaymen'. Occupational identity was perceived as rooted in the collective, even though attacks on it were felt as immediate and personal. It is worth reiterating here the emotional investment made in work by railway employees of this period. Outsiders lacked legitimacy in the eyes of established staff because of their perceived lack of knowledge about the industry and doubts as to their long-term commitment to it.

We now turn to analyse the effect that the changes discussed here had on identity and status. We do so by broadening our focus to include accounts from a variety of commentators.

Nationalisation, modernisation, closure and the nostalgia for permanence at work

Perhaps one of the best ways to understand what was happening to occupational identity in this period, and the way it has been interpreted subsequently, is to see it as not only the active creation of the workers themselves but also as being reflected by a variety of other commentators. Critically, what occurs during this postwar period is that the decline in the industry's fortunes is mapped onto the workforce in a variety of complex and often contradictory ways. Railway employees themselves become responsible for this decline as part of this process, and for some writers there is a deep sense of loss in the very character or calibre of the collective worker. Railway workers are not unique in this respect. One could point for instance to Theo Nichols' (1986) *The British Worker Question* for a general sense of decline, but the rail industry is noteworthy for the sheer amount of nostalgia for lost character within a particular workforce.

This sense of loss is captured beautifully in the 1952 Ealing Studios film *The Titfield Thunderbolt*, which was discussed briefly in the previous chapter (see also Perry, 1981; Barr, 1993; Castens, 2000). The film is important for a number of reasons, but here it is useful in gauging early opinion about nationalisation and its consequences. Titfield can be read as a battlefield on which the three themes of State ownership, modernisation and rationalisation are fought out. The underlying storyline in the film is that the closure of this uneconomic country branch line has been deemed necessary by modernising bureaucrats in Whitehall, obsessed with efficiency and having scant regard for the needs of the local, the small scale and the rural. Closure is proclaimed on the grounds of productivity; opposition to the plan stresses romance, individuality and character against the uniform, the rational and the soulless. The story was a classic Ealing theme: resistance to a remote bureaucracy, stressing 'small is beautiful and big is bad'.

For our purposes in this chapter, though, what is more important is the role played by labour in the film generally, and in particular during the pivotal scene at the public enquiry. Here a range of vested interests – the rival bus company, Whitehall technocrats and a representative of organised labour – oppose the plan of the Titfield amateurs, who want to run the line rather than see it close. Coggett, of the 'National Association of Railway Workers' (who speaks with a northern working-class accent) is portrayed as a union 'jobs-worth' and makes his objection to the plan on the grounds of guarding against 'exploitation' of the volunteer workers, and significantly couches his objection in the formal language of industrial relations:

> My Association will take a grave view of the proposal to employ staff on this line in flagrant disregard of the scale of wages laid down for railway workers.

In response to the vicar's claim that the would-be unpaid railway workers *want* to be exploited, Coggett replies:

> It doesn't matter what you want brother. It's what the bosses want that we are out to stop.

In reply to this, the young squire, who eventually becomes the guard, says:

> In our company there's no quarrel at all between capital and labour.

Thus the union is portrayed as an ideological imposition on the industry, which it is using simply as a vehicle for an irrelevant class war.

The film also takes a sideswipe at the tight labour market that was increasingly a feature of the industry, by highlighting the Titfield company's inability to recruit 'professional' labour. The issue of railway labour itself is notable on one level by its absence, the very inability to recruit a paid workforce serving to emphasise the supposed lack of commitment contemporary workers had for the industry, having been lured away by the promise of higher wages in other employment. In answer to the union official's accusation of exploitation, the squire passionately demands:

> Tell us where we can get some railway workers and we will use them with pleasure: at full union rates!

Lack of labour, therefore, acts as a metaphor for a loss of vocational commitment. The Titfield lobbyists make a plea to be left alone by the unholy alliance of interfering Whitehall bureaucrats and trade union rabble-rousers. The railway can be left in its timeless and self-sufficient world, safe and secure in the hands of the squire, the vicar and an assortment of other local stereotypes who eventually succeed in operating the line at a profit.

For some commentators, therefore, decline is implicitly the result of State control of the railways after 1948. Ironically, it is the very efficiency of the State and the postwar will to enact change, to challenge the old, that these critics object to. A sense of loss permeates these accounts, with nationalisation usually taken as the watershed between a timeless golden age of labour and one dominated by instrumental materialism and modernising zeal. One of the best illustrations of this coupling of structural change, nationalisation and the decline in the moral order of the workplace can be found in George Behrend's elegy to the Great Western Railway, *Gone with Regret*, first published in 1964, some sixteen years after the demise of the private company (Behrend, 1969). In his writings, the Marlborough- and Oxford-educated writer pays homage to the company that a reviewer from *Country Life* (quoted on the back cover) describes as 'that earth-goddess among railways'. He does this by way of accounts of journeys made 'before the disaster of 1947' (Behrend, 1969: 10), which highlight the organic, settled qualities of the past. Indeed the whole book could be read as an attack on State ownership, at times covert, while at others more overt:

> This book is non-political, but if I had a vote, it would not be for Labour, because they destroyed the GWR, which largely helped to bring the railways into their present mess.
>
> (Behrend, 1969: 10)

Throughout *Gone with Regret*, Behrend makes the link between the 'old company' and the spirit of its workforce:

> Much ill-used and often underpaid, these devoted people, whose tolerance and patience at all levels seems wellnigh inexhaustible, for me are epitomized in the person of an elderly porter, who, in the good old days before the period of this book, faithfully served the Cambrian Railways.
>
> (Behrend, 1969: 11)

The narrative employed here suggests that the values of the GWR have survived within a diminishing band of workers who were lucky enough to begin their careers before 1948. Linkage is made between the organisation and the type of worker it bred, with phrases such as 'special', 'craft' and 'fine art' regularly employed in order to mark out the calibre of the former workers from those who followed them. Interestingly there is a sense of lost innocence in such accounts, with nationalisation corroding pride in craft, work and of vocation. State ownership is portrayed as an alien and ultimately destructive imposition into an ordered and hitherto happy world. At one point in the book, Behrend describes a Sunday morning trip to Paddington during the 1960s, and the timeless quality to be found there:

> it could have been 1931! There was not a diesel in the place, and even the West Indians who seem to operate most of the stations today were strangely absent.
>
> (Behrend, 1969: 42)

Importantly we see here the linkage of the 'Golden age' with an essentially White male workforce. Decline in this and in other accounts is associated with the arrival of immigrant labour. Thus the lost world is a white, male, ordered one. The recruitment of non-white New Commonwealth labour is interpreted as an instance of dilution, an erosion of character.

The theme of decline and loss of status haunts the social historical and sociological literature on railway workers in general, and those on the footplate grade in particular (Salaman, 1969, 1974; Hollowell, 1975; McKenna, 1980; Groome, 1986; Wilson, 1996). In all of these accounts this decline in status is explained by reference to State ownership, modernisation – particularly the end of steam – or rationalisation. Some, such as Hollowell (1975), see the initial decline in fortunes as starting in the inter-war period, which was characterised by static promotion and recruitment, with even occupational demotion not being uncommon. Frank McKenna, discussing railway workers as a whole, believes nationalisation was *the* critical point:

> By 1945 the status of railwaymen had gone, and for many railway work was the last resort ... Nationalisation created no new loyalties in the industry. The state and its representatives appeared as remote as the private managers had been in attendance. The railway workers who retired in the 1960s had seen the best and the worst of private enterprise, and they went out of the industry with the knowledge that no industry before or since had engendered and fostered the corporate spirit which characterized the private railway

companies and marked them off so clearly from other complex organisations.

(McKenna, 1980: 63)

And McKenna goes on to lament this loss:

These men had lost something and they never retrieved it. British Rail has failed to give its workforce a specific, corporate identity.

(McKenna, 1980: 64)

State ownership is viewed as an impoverished structure from which to nurture a work culture, and again contemporary railway staff are viewed as less characterful, somehow less authentic than their predecessors.

In Salaman's analysis, changing status is linked to a variety of factors:

The railwaymen said two things were responsible for this decline in status level: the declining importance of the railways as a form of transport, and the decline in the relative economic position of rail-waymen vis-à-vis other working-class jobs. It was also apparent that because of changes in the nature of railway work itself – its organisa-tion, security and technology – the railwaymen too felt that it was very much less of a 'good' job, i.e. one that people admired and wished to have.

(Salaman, 1974: 76–7)

For other writers it was the ending of steam traction and the perceived or actual loss of skill that was the crucial factor in this downward trend. Clive Groome's (1986) book, *The Decline and Fall of the Engine Driver*, is perhaps the best reflection of this interpretation:

The ending of steam can only have come as an anti-climax, the drivers experienced an immediate fall in job satisfaction, no longer were the platforms crowded with enthusiastic onlookers. The 'theatre' of the footplate had ended ... The relative cessation of the need to display skill was noted at once ... The driver now had a 'desk job'. The firemen re-labelled second men, were often removed to a separate roster, so that regular partnership ended. The sense of companionship and loyalty between the two grades was reduced ... glamour was no longer any part of the job.

(Groome, 1986: 57)

The use by Groome of the theatrical metaphor and his description of a blue-collar occupation as 'glamorous' are interesting. He is eliding the public nature of the display of skill or craft with its association with a redundant technology. As the use of steam locomotives finally ended on BR mainlines in 1968, there was a parallel growth in steam preservation and a publishing boom in railway literature that has been criticised by various academics for its overly romantic nostalgia (see for example Kellett, 1969; Strangleman, 2002). In part the end of steam highlighted and intensified the sentimental attachment to the past and the work associated with it. Groome's perspective illustrates the way railway identity was the product of a combination of factors: technology, the workers themselves and their audience.

In many ways this concern with status, mixed with a sense of loss, echoes the sociological discussions of postwar working-class communities and their perceived decline as a result of growing affluence (Goldthorpe *et al.*, 1968). Roberts (1999) has stressed how what he has labelled the 'second wave' of community studies saw the working class as being, or becoming, less morally dense as a result of changing patterns of residency coupled with the material effects of the long boom of the post-war years. The concern over loss can be seen in a variety of works on community at that time and has been an enduring trend in both sociological studies and more popular reportage since. Richard Hoggart's *The Uses of Literacy* (1957) provided a sympathetic portrayal of working-class community that was nonetheless being eroded by cultural change in the period. Less positive visions of such communities can be found in the numerous publications of Jeremy Seabrook (for example 1978, 1982, 1988), which are united by the theme of loss. And this trend resurfaces in Mark Hudson's (1995) portrayal of a former north-east mining community. The recurring theme running through the latter two commentators' work is the idea that the British working class has been bought off or deflected from their historic, sometimes reformist, at other times revolutionary, mission by the relative material affluence of the post-1945 settlement. Thus the depth and dignity of the traditional working class and their communities are exchanged for shallowness and concern for the petty or ephemeral; the high ideals which previous generations struggled for are replaced by the conspicuous consumption of consumer durables.

This similarity between the treatment of occupational and social communities extends to the way (in the latter case) 'positive' aspects of such communities were 'discovered' at the very point they were being destroyed as part of new social housing programmes (see especially

Young and Willmott, 1957). Thus the railway workers, particularly those on the footplate, are of interest to industrial sociologists because they form part of 'traditional proletarian' occupational groups and communities. This sense of an occupation on the edge of losing, or having suffered tremendous loss in, status is reflected in Salaman (1969, 1974) and also in Hollowell (1975), whose research had also been completed during the 1960s. These same groups were seen to be increasingly marginalised as a result of the restructuring of the economy, modernisation of industry and growing affluence, which combined to produce an 'instrumental orientation' to work. Lockwood discusses the trend in his seminal essay 'Sources in variation in working-class images of society':

> Although in terms of social imagery and political outlook the pro-letarian and deferential traditionalists are far removed from one another, they nevertheless do have some characteristics in common. They are first of all traditionalists in the sense that both types are to be found in industries and communities which, to an ever-increasing extent, are backwaters of national industrial and urban development. The sorts of industries which employ deferential and proletarian workers are declining relatively to more modern industries.
> (Lockwood, 1975: 20; my emphasis)

The loss, therefore, is interpreted as the ending of the deep identification with work and the changing nature of the culture of work itself. The railway worker represents the traditional proletarian to Lockwood's (1975) 'privatised' or Goldthorpe *et al.*'s (1968) 'affluent worker' (see also Bulmer, 1975; Brown, 1992: ch. 4).

Thus far this chapter has examined issues surrounding the trans-formation in ownership and technical change with their associated implications for skill level, and the experience of the Beeching era and closure. We now turn to the collective impact these factors had on the status of railway work and occupational identity, understood sub-jectively and intersubjectively.

The loss of status?

One of the greatest changes with regard to the status of railway work during this period was the decline of pay and conditions. As seen in Chapter 2, employment in the industry was highly sought after in the nineteenth and early twentieth centuries because of the security it offered, along with the opportunities for promotion (Kingsford, 1970; McKenna, 1976, 1980; Revill, 1989). Such privilege was undermined in the interwar period as a result of economic recession and corporate restructuring in the wake of the amalgamation of 1923. The trend was temporarily reversed in wartime but the decline in conditions and employment levels continued after the war. This was in a period when unskilled or semiskilled jobs were in plentiful supply in many parts of the country because of the favourable economic conditions of the day (Aaronovitch *et al.*, 1981; Leys, 1989; Smith, 1989). As Howell (1999: 352) points out:

> Previously, the negative features of railway work had been balanced for many by one factor – security. But in a full employment economy that advantage had gone. The renewed resistance to wage increases spelt an early disillusion with public ownership.

After nationalisation pay differentials within the industry narrowed, while simultaneously wages in comparison to outside industry fell in real terms (Gourvish, 1986: 218). At the level of the workforce this contrast is reflected in many of the published autobiographies of the time, as well as in interviews. Several of the accounts written by former railway workers juxtapose their absolute love for the job with the poverty of the wages paid in comparison with other work available locally. Gasson (1981: 117) for example gives a vivid illustration of the dilemma he faced:

> Bill Prior [another signalman] ... asked me if I was aware of the opportunities only a few miles away at the Morris Motors car factory. It seemed from his information that his son-in-law had joined the security staff there working the same hours as we were doing but for five pounds a week more, plus a pension and pay, when off sick ... I wrote to the factory for an interview and a couple of weeks later I was called to attend, and found everything that I had been told was true. I left that interview to discuss things with my wife, and although she knew that the railway was my life, that extra money and the conditions that went with it, were an opportunity not to be missed.

Employment in the car industry is also mentioned in Bradshaw's reminiscences, and again the disparity in wages and conditions is marked:

> The railway fraternity soon learnt that substantially higher wages were now earned by unskilled labour in the new Pressed-Steel

factory producing car bodies for the Morris Motor Company. To achieve a gross income of £12 a week, a signalman would need to work, in addition to his standard 42hrs, a 12hr Sunday and several hours of overtime, whereas in this new factory, it was possible to take home £15–£20 for a 40hr week, and have immunity from either responsibility or unsociable hours.

(Bradshaw, 1993: 169)

In addition to the resentment at the discrepancy over pay and levels of responsibility, there was also the question of long-term job security, an issue intensified both by modernisation and rationalisation:

Burrowing deep into this festering mass of disillusionment was added fuel of doubt and fear, fed by stories of the impending closure of many lines ... Finally, there rose the eternal question: when, if ever, would a pension scheme be forthcoming? ... In those early months of 1959 I recall no less than six local signalmen, all possessing nothing other than railway operating experience, leaving the service voluntarily to seek new careers.

(Bradshaw, 1993: 169–70)

This increased turnover of staff was a problem for railway management in urban centres, where labour markets could be particularly tight. It was a chronic problem in London and the south-east. Bagwell (1982: 281–90) highlights the problems that London Transport faced in terms of recruitment in 1973, when the number of unemployed in the region fell to 1.4 per cent of the workforce. Simultaneously the number of unfilled vacancies more than doubled from 99,200 to 202,200. Several of my interviewees who were employed on London Underground during the late 1960s and early 1970s remember the problems such figures presented at the level of the shop floor. One spoke of the way even very senior staff were leaving the railway with 30 years' service for better paid jobs elsewhere. Others told of the number of cancellations made because of inadequate numbers of train crew. Harry Friend, himself a driver in the Newcastle area, discusses the situation in London:

Down at Old Oak Common [*engine shed outside Paddington*] drivers, never mind about firemen, went to Mars bars factory [*in Slough*] because they got far more money at Mars bars, 'cause Mars bars wanted good workers and they knew they would get them off the railway, particularly off a loco man.

(Harry Friend, interviewed 1995)

One senses in this and other similar interviews an almost ironic pride in the employability of railway workers in outside industry during this period. The implication was that non-railway management recognised the 'traditional' qualities that these workers as a group supposedly possessed, while simultaneously management within the industry actively derided such identity (see Gasson, 1981; Stewart, 1982; Vaughan, 1984, 1987).

There is an interesting regional variation in the experience of alternatives to railway work. For the interviewees from the north-east, changing jobs in this period would have been less attractive than for those of comparable age and grade in the south. Thus there are accounts of increased turnover in lower parts of different grades. This churn in staff also had implications for the status of railway work in two ways. Firstly, it further undermined the image of secure, continuous employment as people left the industry, and BR had to be less selective in its recruitment than had previously been the case. Michael Bonavia, a former chief officer with the BRB and later a respected writer on railway matters, combines sympathy for nationalisation with ideas of decline in the calibre of staff recruited. Writing in 1985 on 'Tomorrow's railwayman', Bonavia is inevitably drawn to the past:

Railway wages failed for a time to keep pace with the general advance; in consequence, recruitment difficulties had sometimes to be solved by accepting unsuitable applicants who before the war would have been turned away. Older railwaymen often regarded them with dismay.

(Bonavia, 1985: 153)

And he goes on to note what this has bequeathed the industry:

Today's railwaymen are therefore a mixed lot. There is a leavening of older men of the traditional, responsible type – good railwaymen in every sense. There are some promising men also. In between come some excellent people, but also some whose attitude to railway work is not all that is should be, who were all that could be taken on in the big cities in the days of over-full employment. They will not leave, now that the economic climate has turned colder.

(Bonavia, 1985: 153; my emphasis)

Note the way urban recruitment is blamed for decline. It is almost as if the body of the railway workforce has been infected by disease. Raymond Williams (1973) reflects on the regularity with which, in portrayals of the country and the city, the purity and innocence of the countryside are compared and contrasted with the disorder and chaos of the urban. At times, he argues, these attributes are unproblematically assigned to the residents of each settlement, and it would seem a similar process is occurring in some of the commentaries explored in this chapter.

The second implication of the large turnover in staff was an undermining of status in terms of a reduction in the seniority needed to fill particular posts, thus upsetting an important moral order in the workplace. In many ways this process had started during World War II with rapid promotion, especially in the locomotive departments, in order to cope with the demands of traffic levels at the time. In the signalling grades the situation could be even more dramatic as younger, less experienced workers took positions that had previously been the preserve of senior employees. Bradshaw (1993: 60) gives his own experience of the resistance he met in wartime:

> I was appearing in the Hunts Bank Offices at Manchester Victoria, facing a Chief Inspector who must have given me just about every possible emergency situation to deal with ... [*He*] let me know in no uncertain terms that I was the youngest trainee signalman he had ever examined; also stressing that, up to the present time, a man had been extremely fortunate if he was able to become a signalman under the age of 30, let alone 20! He didn't know what things were coming to and had no doubt that the consequences of 'boys' of this age being given command of a signalbox would be responsible for many grave accidents in the future.

Vaughan (1984: 116) gives a very similar account of his experiences in the late 1950s, where it was the competition from other forms of employment that was the cause of rapid promotion:

> like most boxes in the area, no one else applied and I got the job. This caused a few eyebrows to be raised and Harry Strong [an older signalman] told me that I was going into a signal box after only nine months on the railway whereas, before the war, men waited nine years for a tiny ... box.

As was the case with Bradshaw, the signalling inspector was concerned about this development and tried, unsuccessfully, to block the move. It is interesting that it is also at the supervisory and management levels that such rapid promotion is seen as a problem, as well as at the shop floor. Again this would seem to point to the shared moral responsibility felt on both sides of the employment relationship towards the industry. It also highlights the ways in which formal functions of the promotional system become codified in work cultures to form expected norms – thus it is assumed that a worker will have a certain amount of experience before being allowed to do a particular job.

From my own experience working on London Transport, there could be periods when promotions occurred very quickly after many years of stability. For example, I gained a position as a Metropolitan Line relief signalman with only four years' seniority, whereas I had been a box boy to signalmen in the same position who had had to wait perhaps ten to fifteen years. A signalman with whom I worked gained promotion to a senior signalling post after a decade-and-a-half on the job, and was referred to as the 'boy'. Many of the other workers in his job were in their fifties or sixties, and had had to wait twenty or more years before gaining this highly sought-after post.

Status decline, as we saw above, was also experienced by the loco men as a result of the end of steam traction. This seems to have been due to the changed intersubjective perception of other grades, supervisors and those outside the industry, rather than simply the objective deskilling of the labour process. Groome (1986) talks about the loss of respect from supervisors and other railway staff. One quotation from a driver on the Southern Region of BR stresses the problem in terms of age and generations:

> In the old days they knew that the driver had done 15 years on the shovel before he passed out for driving ... By the time he was in the top link say at Nine Elms [a shed in South London], they knew the man at the front had done 20 to 30 years service, they couldn't be treated as boys ... Nowadays you can have a man coming on the job ... in five years he is doing top rank mainline trains ... The man on the platform knows they've only been on the job five years and they treat him with contempt.

> (Groome, 1986: 68)

This linkage between the transition from steam to diesel and changing age profiles of footplate staff is echoed in many other accounts. Studs

Terkel (1990) interviewed a retired US railroad engineer (driver) who made very similar comparisons about the moral fibre of the younger workers. The same interviewee also spoke of the difference in supervision:

> The old days, when you had engineer, he was the boss. He was respected as a man and his judgement was respected from the top of the ladder to the bottom. That's gone now. They can get an eighteen-year-old kid out of high school and make him a train master, and you try to tell him right from wrong, he's liable to have you up for insubordination. In the old days, you had judgement on your trains and what you could do. When you figured you had too much, you'd tell the train dispatcher. Your word was law. Respect's lost.
>
> (Terkel, 1990: 561)

Several of my interviewees noted the way the transition from steam to diesel traction brought with it a loss of respect. For example:

> you got this attitude came from office staff and management that anyone could do it.
>
> (George Deownly, interviewed 1994)

Morris Mowbray made a related observation:

> The driver's job really went out of the window when the steam engine went. The driver had a lot of authority in them days ... he was the man; he was like the captain of the ship, you might say, in steam days. But once the steam engine went, basically his authority and everything really went with it.
>
> (Morris Mowbray, interviewed 1994)

Salaman's (1974) discussion of status mirrors this view, but there is also a link with a sense, identified by several of his interviewees, of moral decline in the collective worker within the body of railway workers:

> 'When I joined the railways, a railwayman was king of the working class; now he's a laughing stock. From top to bottom in a lifetime!'

> 'When we first joined the railways forty years ago only the cream could get a job on the railways because it was the best job going.

Now it's all different, now it's the left-overs, the scum that joins the railways.'

(Salaman, 1974: 76)

These kinds of attitudes are also reflected in Hollowell's work. One of his interviewees links decline to immigrant labour being employed by BR (Hollowell, 1975: 231), again echoing the remarks of Behrend recorded previously. What all these accounts indicate is that the status of railway work had declined and that somehow this was implicated in the identity of the collective worker. Perhaps the most important factors were the related conceptions of age and generation. The notion of decline was being coupled with the type of workers recruited to the industry since its 'golden age'. This decline as we have seen was multidimensional, complex and often contradictory. At the phenomenological level it was experienced as a very immediate and personal thing, but this is not to say that the experience of erosion of status was felt at the same time or in the same way by all workers in the industry. In certain grades, the signal workers being a good example, tight labour markets in the post-war period witnessed a more rapid turnover of staff. Decline in these circumstances was connected with the arrival of newcomers.

In the north-east, where alternatives to railway work were less apparent, decline was experienced to a greater extent in terms of closure and the subsequent marginalisation of the industry. This regional aspect of decline is important. In the interviews I carried out, the retired workers seldom criticised younger workers as not being 'proper' railwaymen, as was the case in the earlier works of Hollowell (1975), Salaman (1974) and Groome (1986). In part, this was because the demographic features of the north-east seem to be such that the firemen and later second men with whom the interviewees had worked had themselves been employed in the industry for some time by the stage that decline was apparent. Many of the retired drivers had joined the industry just prior to or during World War II and were not made fully fledged drivers until the 1960s. This seems to have been because a large cohort of staff in the locomotive department who were recruited during wartime subsequently stayed on, a feature of the industry that had important implications in the 1980s and 1990s (see Glaister and Travers, 1993: 40). After this large cohort joined the industry in the 1940s and early 1950s, recruitment to the footplate was less common as staff levels were allowed to drop by natural wastage in the face of traffic decline, closure and technical change. Reluctance to engage in the sort of criticism of younger railway workers found by other sociologists may also have been a function of hindsight, as many of the older workers questioned had been retired for five or even ten years by the time the interviews took place.

For sociologists, especially those researching in the south-east, such accounts of decline seem to have been interpreted as a universalised experience, rather than that of a particular cohort of drivers, at a particular point in their life cycle, during singular times. At the end of Semmens' biography of Bill Hoole, for example, a Kings Cross official is quoted as saying at Hoole's retirement in 1959: 'There are no characters left now, only drivers' (Semmens, 1966: 190). This was nearly a decade before the end of steam. Similarly, one of the men Salaman interviewed in the mid-1960s stated: 'The railways are finished now ... We were the last generation to know what the railways were really like' (Salaman, 1974: 72).

Such readings produce a static account both of the past and the future in relation to occupational identity. The high status is assumed to have been a permanent fixture of such identity in the past, rather than one that was open to negotiation and modification under pressure as part of the routine movements of the relative strength of capital and labour over time. While it is true that there was much stability in the industry, particularly amongst certain grades, there was also always movement and change (Savage, 1998). One only need look in the retirement notices of the staff magazines of the early decades of the twentieth century to see the rapid promotion that had been a feature of the industry in its expansive phase in the previous century. During the formative years of the industry, promotion to driver might occur after only a couple of years, whereas in the 1930s a fireman might have to wait nearly three decades for the same post (see Chapter 2; see also Kingsford, 1970; McKenna, 1976, 1980; Revill, 1989). In many accounts this flux is lost from view as authors engage in a discussion of comparative statics. By this, I mean positive views of the past are juxtaposed with negative ones of the present (see Salaman, 1969, 1974; Hollowell, 1975; Groome, 1986).

Railway workers are far from unique in being seen as having suffered from, or being on the cusp of, loss. There is of course a long tradition of perceiving work as having lost meaning, or contemporary workers (in whatever period) as being less committed or characterful. Theo Nichols (1986: 6) reminds us that among right-wing commentators:

the ideological castigation of the workers can be traced back over a very long period indeed – right from the onset of the Industrial

Revolution; right through the nineteenth century ... and throughout the eight decades of this century as well. As far as the British
workers and trade unions go, it has also been a recurrent theme that
things are getting worse.

This sense of loss has also been a feature of sociological discussions of
work and community reaching back to Marx, Durkheim, Weber and
Tönnies, with modern capitalist society being marked by alienating,
estranging but ultimately necessary forms of employment (Stafford,
1989; Turner, 1987). And Turner argues that each of the founding
fathers tells us something about a nostalgia for a lost social stability.

This concern could be interpreted as the nostalgia for permanence, the
search for stability in work, the sense in which permanence is 'good', and
that modernity has somehow disrupted an equilibrium that should
ideally be restored. David Meakin (1976), in his book on literature and
work, devotes a whole chapter to this notion of nostalgia for permanence
in work. He views it as a recurring theme in the writings of, amongst
others, George Eliot in *Adam Bede*, the Wessex novels of Thomas Hardy
and Tolstoy's *Resurrection*. Meakin argues that such novels reflect a profound respect for skilled craftwork or, more generally, labour rooted
within small communities and imbued with a timeless quality. It is precisely this sense of permanence and timelessness that is important. The
stability and ordered predictability of traditional labour is juxtaposed
with the rootless impermanence of 'modern' employment. Modern
industry predicated on urban settlement is interpreted here as:

> uprooting individuals from the old traditions that have given their
> ancestors stability and a sense of gravity in the world.
>
> (Meakin, 1976: 61)

There is a mixing here of several senses of fixity and instability –
spatial, social and occupational are all implicated in the understanding
of transition and the concomitant feeling of loss. Important also is the
sense in which traditional, more established types of work are threatened by the onward march of modernity. The authors' characters
inhabit a border country, to use Williams' (1973) phrase, between tradition and modernity.

As we have seen above, this nostalgia for permanence at work is not
restricted to pre-industrial work. The 'affluence' debate of the 1960s
could be interpreted as a lamentation for 'traditional' industrial work.
Predating this academic charting of loss, left-wing British film-makers

in the 1930s and 1940s, influenced by European socialist realism, attempted to capture the 'dignified labour' of steel workers, shipbuilders and coal miners. The definition of 'real work' is widened here from craftwork in a rural setting to mass production in the urban. Paul Swann, in his fascinating account of the Documentary Film Movement (DFM), quotes Edgar Anstey, an influential member of this group, on his attitude to the representation of manual labour:

> the workingman can only be a heroic figure. If he's not heroic, he can't be a workingman.
>
> (Swann, 1989: 88)

Whilst this celebration of industrial work marks a shift from accounts that see legitimate work as only pre-industrial and small scale, it nonetheless shares a sense of impending loss, of heroic employment being subsumed by the mundane. This discovery of value at the point of loss echoes Raymond Williams in *The Country and the City* (1973), in which the continual tension between the urban and the rural is apparent. In particular, Williams reflects on the idea of a lost 'golden age' which is always seen as having *just* disappeared within living memory – 'Just back, we can see, over the last hill' (Williams, 1973: 9). He notes this is a problem of perspective, and he likens this constant appeal to a lost arcadia to being on a moving escalator casting a backwards glance. He goes on to stress the importance of recognising what this continuity within accounts of the recent past tells us about the structure of feeling in those who produce it. For us, then, what does this structure of feeling that constantly harks back to a golden age of railway work and workers tell us about the industry and employment in it?

While there are important similarities between the reflections on the past of those inside as well as outside the railway industry, it is possible to argue that there are equally important differences. For those outsiders who commentate on railway employment there is a shared sense in which a process of the abstraction of labour, or occupational identity, is occurring. Certain characteristics are attributed to, or idealised around, stereotypical workers. As in any case of stereotyping, the empirical reality within a structural context is somewhat different, and more complex. Academics are similarly guilty of such stereotyping, when in seeking out the traditional worker they find only traces of such an identity. Thus in their view the real, authentic, heroic worker is always just out of reach, while the contemporary worker fails to live up to the standard set by previous generations.

Workers' accounts of the past need to be treated slightly differently. Analytically Fred Davis's (1979) work on nostalgia is important here. In *Yearning for Yesterday*, Davis identifies three distinct levels at which nostalgia works: simple, reflective and interpretive. At the *simple* level, nostalgia is the largely unexamined belief that 'things were simply better in the past'. In *reflective* nostalgia, people do more than sentimentalise about the past, they begin to raise questions about truth claims. Finally, in the realm of *interpretive* nostalgia, the emotion itself is rendered problematic. The actor will seek to objectify the nostalgia felt. For the workers at the shop floor, the experience of change is necessarily complex and multifaceted. In most cases, workers' accounts avoid 'simple' nostalgic accounts of the past, but are instead reflective or even interpretive in character. For this group, choices made and action taken are situated within a nexus of competing pressures.

Perhaps an even more developed sociological analysis of workplace nostalgia, or more accurately the interpretation of the past, can be found in the neglected work of Peter Marris. In his book, *Loss and Change* (1974), Marris draws parallels between seemingly disparate examples of social change from his research. He sees a linkage between them in terms of the way loss and change are coped with. Fundamentally his thesis is that nostalgia or attachment to the past and the perceived conservatism that such attachment implied are necessary if individuals, groups and whole societies are to change and evolve. The conservative impulse, far from blocking change, actually facilitates it by sanctioning development based on past experience. Nostalgia for Marris can therefore be interpreted as a kind of bereavement process, the working out of grief marking the end of previous attachments.

These accounts give us access to the experience of decline or change at the phenomenological level of the workforce. Interviews and autobiography are snapshots of attitudes to a changing process, rather than idealisations of a fixed past as perhaps is the case in some of the other reflections. In this way we can understand both change *and* the continued appeal of the recent past as a golden age. In the following chapters I will return to these issues of loss and change and the ideas of Marris and Davis, as well as those of other theorists.

Conclusion

The period 1948–79 was marked by radical change – wholesale rationalisation and closure, massive job reduction and a decline in the collective status of railway employment both for the workforce itself but also for outside commentators. The writing about, and the study of, work in the railway industry during this time are imbued with a deep sense of loss, with workers linked very directly to the decline in the industry's fortunes. For some academics and those on the Left, the period was one in which, as in other industries, commitment to work, craft and vocation was being undermined by a series of linked societal shifts in the norms and values of the working class. Railway work and the positive aspects of occupational pride were being eroded, replaced by a new type of worker less concerned with the intrinsic value of work itself than with monetary reward. Right-wing commentators also view the status and standing of railway work as declining during this period. State ownership had undermined worker commitment and identification with work, a situation made worse by the power of trade unions and their insertion of the cash nexus into the employment relationship.

Two key points emerge from this chapter to be taken forward. Firstly, there remained within the workforce a strong sense of occupational identity, and this 'railroad culture', as some have labelled it, ensured that the tremendous change witnessed during this period was both rendered understandable and coped with fairly successfully by those on the shop floor. Secondly, the image of work in general, but railway work in particular, has important implications for the way in which the industry was treated during the 18 years of Conservative rule from 1979.

5

Back to the Future? Railway Commercialisation and Privatisation, 1979–2001

> As if shamefaced about the boldness of its new design, Liverpool Street Station has just added ornamental gates with the words 'Great Eastern Railway' inscribed in a cast-iron, multi-coloured medallion. It is as though the station were turning its back on not only half a century of BR (the 'British Railways' of nationalization) but also on the amalgamation of 1922 which produced the LNER.
>
> (Raphael Samuel, *Theatres of Memory, Volume 1: Past and Present in Contemporary Culture*, 1994: 115)

Introduction

If, as L.P. Hartley (1958) remarked, the past is a foreign country, then the last two or three decades have seen the development of mass tourism amongst the British political class. While 1979 has been viewed as a major watershed in British political, social and economic life, it has not prevented the near constant plundering of the past as either a warning or a guide for future conduct, for in that past there is much that is useful. The collectivism of the era of political consensus, and in particular the policy of nationalisation, have been carefully nurtured as icons of failure by neoliberals in both the Conservative Party and most recently New Labour. The period from 1945 is portrayed as an aberration, an era in which the country lost its way, when managers were cowed by overpowerful trade union dinosaurs, when individuality was crushed by collective provision, and imagination and creativity were held back by the inertia of bureaucracy and red tape. However, the past has also been drawn on as offering a model for future behaviour. In this reading of history, change is necessary in order to return

104

to a utopian past, a 'golden age' marked by entrepreneurial endeavour, when workers were loyal, dependable and committed, and finally when the market was the organising principle behind all aspects of life.

Nowhere has this debate over the meaning of the past been more apparent than in the railway industry since 1979. This chapter examines the commercialisation and subsequent privatisation of the industry through the distinct but related themes of culture, change and nostalgia. These themes allow us to explore the way the past, in both utopian and dystopian readings, is used to construct the present as unsustainable or as being in crisis and therefore in need of change. But the menu of possible directions of that change is limited to solutions based on a simplistic, narrow and overly idealised view of the market.

Organisational culture

As we saw in Chapter 1, the 1980s and 1990s were marked in managerial terms at least by a focus on questions of culture. In the social sciences, culture has been notoriously difficult to define (Williams, 1983; Eagleton, 2000; Smith, 2000). In management writing however, this ambiguity allows the term to become a highly useful catch-all phrase to characterise the 'people issues' in organisations (Grint, 1995; Parker, 2000). The organisational anthropologist Susan Wright (1994: 4) noted the way managerial commentators and writers used notions of culture:

> Culture has turned from being something an organisation is into something an organisation has, and from being a process embedded in context to an objectified tool of management control. The use of the term culture itself becomes ideological.

For Wright and other critical social scientists this use of the term culture was alien and profoundly disturbing. What they objected to was the crude way in which the complexity of the term employed by sociologists and anthropologists was ignored in order that it could be used in management writing and practice. The management writers shared an understanding that organisational culture *could* and *should* be manipulated by managers in order to achieve corporate success. One of the most famous of this genre of 'culture change' writing is Peters and Waterman's *In Search of Excellence* (1982), the central theme of which was that the truly successful organisation possessed the 'right' culture. The following quotation from Deal and Kennedy (1988: 15) is typical of this use of the concept:

> We think that people are a company's greatest resource, and the way to manage them is not directly by computer reports, but by the subtle cues of a culture. A strong culture is a powerful lever for guiding behaviour; it helps employees do their jobs a little better.

This notion of 'strong culture' gives a clue to both the concept's attraction for managerial writers and why it is so problematic for more critical writers. Inherent in such an understanding is that strong cultures are desirable and, equally, weak cultures are to be avoided. There is an implicit reliance here on overly simplistic structural-functionalist understandings of social organisation, with the taken-for-granted assumptions of normal and deviant behaviour. Lynn-Meek (1988) believes that the attraction of this approach lies in the embedded biological metaphor. Healthy organisations are those where deviance is marginalised. It is a short step to a position where an essentially unitarist conceptualisation of the organisation is the assumed norm, or at least the aim. In this reading, management take on the role of the final arbiter as to what *the culture* should, and more importantly perhaps, should not, be. At the heart of such a mobilisation of culture is a model of history, a sense in which certain aspects of the past are legitimate, or illegitimate, and that further change is necessary and desirable if an organisation is to succeed. Discussions of culture and its manipulation are therefore inherently historical in their approach. Such programmes can be thought of as an exercise in the delegitimisation of the past, or a particular interpretation of the past, and the most powerful aspect of such a process is the attempted isolation of those who seek an alternative vision of the past. Often the attachment to the past is labelled nostalgic in a highly simplistic and pejorative sense. It is this aspect of the culture change initiative that we go on to examine, both in this chapter and in the following one.

Nostalgia

Yiannis Gabriel's (1993) important contribution to the debate on nostalgia in organisations was part of a larger collection on emotion in organisations (Fineman, 1993), which itself could be seen as part of a wider renewal of interest in aspects of subjectivity at work (see for example Grey, 1994; Casey, 1995; Sosteric, 1996; Albrow, 1997). Gabriel suggests that gauging sentimental attachment to the past within the workplace is fundamental to comprehending the full significance of emotions in general. Mapping out the ground on which organisational nostalgia could be studied, he argued for:

> a view of nostalgia as encompassing a range of distinct emotional orientations in organizations ... organizational nostalgia is not a marginal phenomenon, but a pervasive one, dominating the outlook of numerous organizational members, and even defining the dominant emotional complexion of some organizations.
>
> (Gabriel, 1993: 119)

The distinction is drawn between two separate areas of the study of nostalgia: the rise of heritage in a public sense and the role of personal attachment to the past. Gabriel draws attention to the use of nostalgia within popular culture, where 'whole sectors of the economy are fuelled by nostalgia' (Gabriel, 1993: 119), and examples are cited from advertising, film, television, music and the arts.

The debate over the development of this 'heritage industry' is itself a lively and fascinating one, with great controversy over the meaning and consequences for a society that increasingly reflects on its own past. On the one hand, there are those who see this trend as a deeply conservative and backward-looking one. Hewison (1987), for example, made a polemical attack on what he labelled 'the heritage industry'. While Hewison saw museums as good in themselves, there was a sense in which quantitatively there was a trend towards the blanket term of 'heritage' being attached to anything 'old', and consequently heritage meant everything and nothing because of its indiscriminate usage. This, he believed, marked the 'imaginative death of the country' and as a consequence, a country like Britain, obsessed with its past, would be unable to face its future (Hewison, 1987: 9).

A more measured and academic treatment of the meaning of contemporary nostalgic fascination with the past has been developed by Patrick Wright (1985, 1993, 1996). In *On Living in an Old Country*, Wright (1985) explores the relationship between the rise of heritage and the backlash against the postwar consensus. He too focuses his attack on the regressive consequences of a backward-facing culture. This concern is echoed by Furedi:

> The sense of the past seems to preside over vast areas of social life. In culture, intellectual life and private hobbies, past times are

impregnated with nostalgia ... The veneration of the past reflects a mood of conservatism.

(Furedi, 1992: vii)

There is, however, a more positive reading of this 'heritage turn' available in the work of authors such as Samuel (1994) and Lowenthal (1998). In both of these, the greater visibility of popular forms of history has a more ambiguous meaning, and in Samuel's work it is viewed as part of a radical widening of history from an elitist and narrow base. What emerges is a society more aware of its own past, more able to question both its origins and its future. Indeed, Samuel (1994: 265) claims that radical 'heritage baiters' are themselves 'pedagogically quite conservative'. Similarly, Davis (1979) argues that the audience for heritage, and nostalgia, does not passively receive a given story about the past but actively engages with it. Indeed, Davis goes further and suggests that nostalgia is *only* possible if based on a personally remembered and experienced past.

A second strand that Gabriel identifies, and which is perhaps more explicitly related to his work, is the social-psychological study of nostalgia, which in extreme cases leads subjects to 'literally live in the past' (Gabriel, 1993: 120). Gabriel argues that organisational theorists have avoided this facet of the concept, and he stresses its importance in comprehending the attitudes, assumptions and actions of organisational members. Crucially, he points out that organisational members often cling doggedly to images of a 'golden past' during or after restructuring and corporate culture change programmes. Here the past is associated with pleasant experiences, the present its pale reflection:

in this sense, nostalgia is a state arising out of present conditions as much as out of the past itself.

(Gabriel, 1993: 121)

Davis (1979) goes further in arguing that nostalgia uses the past, but is not a product of it. He disputes the idea that any era is innately more open to a nostalgic reading than any other, and argues that it is precisely contemporary concerns which lead to a particular annexation of the past.

Munro (1998) discusses the concept of *belonging* with regard to organisational actors' attitudes to the past. 'Belonging' in this sense is the attachment to place or people within the context of a company or department. Again, as in Gabriel's and Davis's understandings of nostalgia, a heightened sense of belonging develops precisely when such identities are challenged. Interestingly, Munro makes little use of the term nostalgia in his work, although he stresses that *belonging* implies a rootedness in the past.

Gabriel also comments on Davis's (1979) deliberately underdeveloped concept of 'nostophobia', viewing it as useful in analysing the way some organisational members reverse the positive/negative association between the past and present. The nostophobic see the past as an era to be escaped from rather than being warmly remembered, while the present and future are more actively embraced. Gabriel develops this concept to explore his own material, arguing that marginalised workers attach nostalgic meaning to a wide variety of animate and inanimate objects in their work lives, and derive from such attachments a sense of ontological security. The past, in this register, becomes a refuge from the flux and general instability of the present. Corporate history, and the individual's place in it, acts as a powerful resource for the sustaining of occupational identity. There are interesting parallels here with Munro's (1998) account of corporate 'rubbishing of the past', where a new management deliberately aims to destabilise an established workforce's conception of the past and hence their sense of belonging to it. It could be argued that the increased interest in nostalgia within the workplace and about work more generally reflects a concern with the speed and scale of change in industrial societies. In particular there would seem to be a welcome concern with the contemporary meaning and future of work (Rifkin, 1996; Milkman, 1997; Sennett, 1998). Such a concern is also echoed in various shop-floor studies of work produced recently (see for example Casey, 1995; du Gay, 1996). In both Casey's and du Gay's work, a distinction is made between generations within the workplaces studied, and both have noted the way older workers looked favourably on the past when compared to the present and future.

While it is important to recognise the value in the work of Gabriel, Davis and Munro, it is possible to build critically on their insights. Firstly, while Gabriel has raised the profile of nostalgia within organisational research, he is overlooking previous contributions to this debate. The issue of nostalgia, along with other emotions, has regularly surfaced within industrial sociology. Two examples are available in the work of Nichols and Beynon (1977) and Alvin Gouldner (1964). Both pieces include discussions of workers' attitudes towards older and newer management, with the latter tending to fail to live up to the previous incumbents' warmly remembered legacy, labelled by Gouldner in *Patterns of Industrial Bureaucracy* (1964) as the 'Rebecca myth'.

A second point about Gabriel's work is the absence of a discussion about how management themselves make use of nostalgia. In his account, sentimental attachment to the past is almost exclusively the province of the shop floor. Similarly, in Munro's writing, management only ever perceive the past as a negative other to the present and future – history is 'resurrected' *only* to be 'rubbished' (Munro, 1998: 221). The argument of this book is that, on the contrary, management and politicians actively use notions of nostalgia and nostophobia in the ideological construction and portrayal of the past with respect to the present and future. It questions Davis's essentially benign interpretation of nostalgia and the idea that such an emotion is always predicated on lived experience. This allows a more sophisticated understanding of the role which history (as discourse), nostalgia and the past play in organisational life.

These themes are considered here in the context of successive Conservative administrations' mobilisation of the 'past' in their treatment of the nationalised railway industry, and the way British Rail and its successors dealt with their history. The past, history and identity are almost infinitely malleable and are used by management and politicians to win consent for change, or at least to marginalise criticism among workers, management and the public. Nostalgia and its opposite are harnessed here in the reworking of organisational ontological security, so that resistance based on attachment to the remembered past is isolated. What is interesting in the context of the railway industry in the UK is the way iconic portrayals of the industry and its workers resurface time and again in both negative and positive portrayals of the past, as we saw in Chapters 3 and 4. In drawing on such imagery, politicians and managers have recognised the cultural currency of these images.

The railway industry 1979–90: the nostophobic years

With the return to power of the Conservative Party in 1979 under Mrs Thatcher, the railway industry in the UK was to suffer in two ways. Firstly, along with other public corporations, the railways were seen as ripe for reform. Peter Parker, then Chairman of the British Rail Board (BRB), wrote in his memoirs of his meeting with Thatcher during the late 1970s: 'to be nationalized, she explained, was an industry's admission of failure', and logically only failures worked in the public sector

(Parker, 1991: 304). The Thatcher project, as in retrospect it has come to be seen, was based on a conception of the economy and polity as being threatened by the unrestricted growth in the size and scope of the State. Thatcher's aims were to liberate the private sector from State restrictions, reduce the size of Government through tight control of public sector borrowing, and rein in organised labour (see Hall and Jacques, 1987; MacInnes, 1987; Smith, 1989; Pollitt, 1993; Hay, 1996). This negative reading of the postwar settlement was developed as the justification for wholesale change. Hay (1996) notes the way the 'crisis' of the 1970s, especially the 'winter of discontent' of 1978–9 and the role of organised labour, was ideologically drawn on in subsequent political discourse in a construction of the 'past' over the next 20 years.

The second problem that the railway industry faced under Mrs Thatcher was that the new administration was overtly hostile to rail per se, representing as it did an inherently collectivist and old-fashioned form of transport. Alan Clark, the right-wing MP, provides in his diary an insight into the Thatcherite attitude to the industry and its workers during the 1980s:

> I am sitting in a first class compartment of the Sandling train, odorous and untidy, which, for reasons as yet undisclosed, and probably never to be disclosed has not yet left Charing Cross. 'Operating Difficulties', I assume, which is BR-speak for some ASLEF slob, having drunk fourteen pints of beer the previous evening, now gone 'sick' and failed to turn up.
>
> (Clark, 1993: 169)

While this may represent the more pugnacious end of Conservative opinion, Clark's general attitude was reflected in more mainstream Tory thought during the 1980s. Bagwell (1996: 120) examined the way the Tories were 'pro-car' and 'anti-railway' during the 1980s, with State support for the railways falling progressively from £1.2 billion in 1982 to £400 million in 1991, while UK expenditure on roads rose from £3.5 billion to £6 billion. Support from public funds for BR was, during the 1980s, the lowest in Europe as a percentage of GDP, falling from a high of 0.3 per cent in 1983 to nearly 0.1 per cent in 1990. The Community of European Railways' average in the same years was 0.7 per cent and just under 0.5 per cent respectively.

This combination of omens was all the more unfortunate given the pressing need for greater levels of investment to replace the worn-out equipment of the Modernisation Plan era, dating back to the mid-1950s (see Chapters 3 and 4; also Gourvish, 1986). Under Parker, the BRB stepped up the search for greater levels of productivity, which they hoped would encourage the Government to sanction more investment. At first, the Board took the railway trade unions with them – Pendleton (1993) describes this period as one of 'tacit alliance', with the unions cooperating in realising productivity gains.

In the early 1980s, however, the BRB produced six key demands on productivity, which affected most notably the train drivers' grade. The one on flexible rostering caused a series of damaging strikes during 1982 (Bagwell, 1984; Parker, 1991; Pendleton, 1991a, 1991b). This was an attempt to move away from the guaranteed eight-hour day, which dated back to 1919 and was an article of faith with ASLEF, the drivers' union (see Raynes, 1921; McKillop, 1950). The BRB aimed to replace the fixed eight-hour shift with a variable period of between seven and nine hours, with the ultimate aim of reducing the number of depots and the overall establishment (Pendleton, 1991a, 1991b). The choice to fight on the ground of the eight-hour day was by no means accidental for either side. Both the BRB and ASLEF knew the historical and emotional significance of the issue. The union saw the agreement as perhaps their greatest triumph – Murphy (1980: 30) describing it as the 'Old Testament' – while for management, it was an anachronistic barrier to *modern* working practices. Parker himself couched his criticism in terms of perceived sentimentality on the part of the union and its members:

> The driver grieved over past glories. Once an engine-driver was what most boys wanted to be; now nobody bothered to stop and have a word with him about the journey, good or bad. Once the King of the Road, rising from cleaner, through fireman, to the throne on the footplate; now he was in unglamorous exile. Once he was inseparable from his locomotive, his castle and his home, cherishing it and its reputation for meeting the timetable, frying breakfast on the shovel; now he and it were computer-programmed through the depot. Once sure of his place, superior, knowing the shedmaster; now he was adrift in the impersonal professional world of area managers and operating managers … Gradually the driver's grip on things was in truth slipping. It was as if esteem was lost with the loss of steam, self-esteem and other people's. And as I was bringing the chips down for change at the beginning of the Eighties, just when the union most required a prophetic, restoring vision, the executive was in the hands of class warriors battling in the trenches

of 1919 agreements which had enshrined the sanctity of the eight-hour day.

(Parker, 1991: 258–9)

Note the way Parker draws on the nostalgia surrounding steam railways, but also blends this with the idea of backwardness associated with the failed military strategy of the Great War. The importance of the issue of flexible rostering for our purpose is the light it sheds on the changing relationship between the BRB, the trade unions and the workforce. The dispute was seen as a symbolic shift away from the previous era of corporatist industrial relations, paralleling similar moves made in other nationalised industries (see Ferner, 1988, 1989; O'Connell Davidson, 1993; Pendleton and Winterton, 1993). In each case, it was the management that cast itself as *modern* and *progressive* and the unions and their members were registered as *inflexible* and *reactive*.

As part of the process of organisational change begun by Parker, a policy of sectorisation was implemented from 1982 (Gourvish, 1990). The aim was to introduce a more market-orientated approach to the railways with the clarification of managerial roles and responsibility, and sectors formed under the headings:

- Intercity
- London and South East Passenger
- Provincial
- Freight
- Parcels.

Sectorisation was a major reorganisation for the industry, and in part was designed to demonstrate to the Government that the railways were being professionally run in spite of the continuing need for subsidy. The process was accelerated after 1982, when the Government set Intercity the target of breaking even by 1985. The Conservatives' aim at this point was not privatisation of the railways, but rather the restructuring of the industry in order to cut subsidy while allowing the sale of profitable parts of the BR empire (Gourvish, 1990; Vincent and Green, 1994; Harris and Godward, 1997). The architect of these changes was Bob Reid, who was to succeed Parker in 1983 as the chairman of the BRB. Ironically Reid, unlike the 'outsider' Parker, was himself a career railwayman, having joined the London North Eastern Railway in 1947 as a traffic apprentice (Gourvish, 1986).

Reid's aim was to liberate and devolve decision-making in the organisation, allowing choices to be made at a more appropriate level, believing that the previous structure had allowed front-line and middle managers to hide behind the historic legacy of bureaucracy and the industrial relations machinery of negotiation. Sectorisation aimed at giving a sector director 'clear bottom-line responsibilities' and, therefore, the ability to act as a purchaser of services from the other parts of the organisation (see Reid in the Foreword to Gourvish, 1986). As part of this process, the area level of the organisation was abolished, these responsibilities being devolved or allocated to a sector or regional tier. The aim, as Bate (1990: 88) has put it, was to shift BR from being 'a "production-led" to a "business-led" organization' – in other words, to a situation where it was 'more responsive and more commercially focused on grasping competitive advantage'.

Reid's contention was that in the past, the industry had been dominated by engineering and operational considerations rather than marketing or commercial ones. Sectorisation aimed at breaking this cycle by inserting a market relationship between supplier (production) and customer (business sector). The breaking of the organisational power of engineering and operational interests was to have a profound effect on the industry and is discussed in subsequent chapters. Munro (1998) talks about the way the 'market' is portrayed in organisational change programmes as 'modern', and that to resist its introduction into any area of corporate existence is to immediately invite the charge of backwardness. This assumption is reflected in the case of BR. Bate argues that Reid identified the industry's culture and history as problematic, and quotes Reid as saying:

> For almost 100 years railways had a monopoly with no competition. Traces of that history still remain in the culture of the industry, which has been slow to believe that its monopoly has been broken … there has been a nostalgic hanging on to history.
>
> (Reid, quoted in Bate, 1990: 86; my emphasis)

In distancing the organisation from its past, Reid, like Parker before him, intended to send a signal not only to his staff but also to Government that things had changed. As Ferner (1988: 65–6) puts it:

> the reform process itself is part of the currency of political debate. Ministers are enabled to defend BR's interests by arguing that its planning is now more 'realistic' and 'professional', rather than

'over-optimistic' ... and the entrepreneurial language of the 'bottom line' and 'value-for-money' enters the processes of political exchange and bargaining between politicians, civil servants and corporate actors.

As part of the process of sectorisation, the position of organised labour and management was challenged, and the battleground was specifically drawn in terms of the 'past' versus the 'future'. This organisational conflict above shop-floor level is illustrated in the following quote from Bate (1995: 150):

British Rail is a good example of this contested cultural terrain. During the 1980s culture and counter-culture fought it out as the old guard in 'Production' clashed head-on with the young turks in the 'Sectors', each parading before the other ideologies and styles of thought which they knew to be provocative and unacceptable. The one side valued service ('value for money'/the social railway), the other side profit ('money for value'/the commercial railway). These were life issues, issues on which no inch of ground could be conceded. As one participant put it – a sector manager in fact – 'there is a clash between those who seek professional management and those who see railways as a quasi-religion – something to which normal commercial disciplines shouldn't or couldn't apply'.

This tension is also echoed in John Storey's (1992) work. Storey quotes a BR regional finance and planning manager who distinguished between the business side – which was seen as filled with 'younger, very good people' who were 'enthusiastic to make change happen ... the production people'. This second group were described as 'older and more established', 'long on experience but reluctant to change' (Storey, 1992: 206). Bate has argued that such was the success of this process that within five years of the introduction of sector management, every one of the former regional general managers in production had moved post, and in most cases this had not been by choice (Bate, 1995: 150–1). This shift towards the dominance of 'business-led' culture culminated in 1991 with the introduction of the 'Organising for Quality' (O for Q) scheme, which further strengthened the sectors. These then became semi-autonomous businesses under the BRB, each having responsibility for their own production, engineering and operational functions (Vincent and Green, 1994; Pendleton, 1995; Harris and Godward, 1997: 54). But this structure only lasted a few years before the wholesale privatisation of the industry took place, sectorisation itself being deemed to have failed the test of responsiveness to the market.

There are interesting similarities and differences here with the corporate restructuring that took place on the London Underground (LU).[1] As on the mainline railway, LU management throughout the 1980s sought cost savings and greater flexibility from its staff through policies such as reductions in establishment (the number of workers employed) and one-person operation of trains. However, during the late 1980s and early 1990s the company enacted a far more ambitious 'culture change' programme called the 'Company Plan'. As Storey and Sisson (1993: 17) noted:

> London Underground's wide-ranging plan consists of a package of measures which include staff reductions, salary status for all employees, the abolition of premium pay for weekend working, a reduction in the number of separate grades of staff from 400 to 70, and the replacement of promotion based on seniority to one based on assessed merit.

This scheme involved reducing staffing levels by a quarter, or some four thousand personnel. Remaining staff were given new contracts of service, with the expectation of greater levels of flexibility. The Company Plan was the brainchild of a US consultancy firm, and during the changeover period the past and future of the organisation were described as 'Old World' and 'New World'. Much of the rhetoric around the Company Plan involved the portrayal of the 'Old World' as being inflexible, unresponsive, bureaucratic and overly rule bound. Those who resisted changes were referred to as 'dinosaurs', their opposition labelled 'nostalgia' in a pejorative sense.

Thus history was defined within BR and LU during the 1980s in a negative, or nostophobic, sense. The past is something to distance oneself from; the future belongs to those who can adapt to the new emphasis on the market and its demands.

Privatisation 1990–8: nostophobia meets nostalgia

Ironically it was the process of privatisation itself that marked a shift from this rather one-dimensional view of the past, as history now potentially offered a template for the present and future. The Railways Act 1993, which led to privatisation, became law on 1 April 1994 (Bagwell, 1996: ch. 13). Throughout the 1980s, railway privatisation had been rejected on the grounds of complexity and a recognition of the need for continuing public subsidy. Even Nicholas Ridley, one of the many Secretaries of State for Transport (1983–6) under Thatcher and a keen advocate of denationalisation himself, advised her that it would be 'a privatisation too far', and later during the debate on privatisation in the Lords he actually spoke against it (Harris and Godward, 1997; Wolmar, 2000). By the early 1990s, however, privatisation even here was made possible by the changed nature of BR as an organisation, the 'success' of other denationalisations, and the political appetite for further privatisation within Conservative ranks (Seldon, 1998).

In the wake of the re-election of the Major Government in April 1992, proposals for privatising the industry were unveiled. Harris and Godward (1997) have argued that the programme was heavily influenced by the ideas of the neoliberal 'think tanks', and believe that the policy, and the details of it, were driven by ideology rather than by experienced railway management (see for example Glaister and Travers, 1993; *Guardian*, 6 November 2000). The reason for this absence of substantive railway industry input is twofold. Firstly, there was Government distrust of those in the industry whose criticism was interpreted as organisational conservatism. Secondly, there was a reluctance on the part of senior managers to fragment the industry along the lines desired by neoliberals on the grounds that it was practicably unworkable. John Welsby, the last Chair of the BRB, rejected the accusation that BR had tried to sabotage the privatisation. Rather, he suggested that Government ministers and their advisors:

> didn't understand the nature of the markets they were dealing with, how to produce train services or what the implications of their decisions were. They had taken no advice from anybody who knew anything about it and the consequence was one of frustration.
>
> (Welsby, quoted in Wolmar, 2000: 128)

John Major had himself mused over the possible shape that the privatised industry might take. In late 1991, according to a later article in the *Guardian* (17 February 1992), 'John Major launched the idea of returning to the "good old days" of the four regional railway companies'. And Seldon (1998: 260) noted in his biography that:

> Major, joined by Patten, wanted privatisation on a regional basis, recreating the sentimental pride of the companies such as the Great Western Railway.

Major's need to differentiate himself from his predecessor may well explain both the policy of privatisation and the imagery drawn on here. In this sense, nostalgia for the interwar period is mixed with an attempted partial annexation of a postwar 'Ealing Studios' view of British, or more precisely English, society (see the critical remarks of Barr, 1993: 182–3). This was the same 'nation at ease with itself', where warm beer could be consumed while watching the village cricket team on the green. Intellectually, the use of such imagery drew on the speeches of Stanley Baldwin (1937) and perhaps, rather ironically, echoes the writing of George Orwell – most notably in his essay *The Lion and the Unicorn* (Orwell 1984). Again, there are interesting similarities with and differences from Margaret Thatcher's use of the past. Unlike Major, Thatcher looked to the Victorian era as her talisman (see Samuel, 1998: 330–48). But, like her successor, she did so selectively, so that the era became, in a sense, out of time and place, the positive aspects of the period being distorted and magnified, the negative ones ignored.

Major's nostalgic view of the interwar period, and in particular its railways, is reflected elsewhere in Conservative political discourse. Bagwell quotes the right-wing patrician MP Sir John Stokes from a Commons debate in 1991 comparing the nationalised industry with the Big Four era:

> I remember the thrill of going to Paddington station en route to Oxford, and I remember the chocolate and cream coaches, the glorious engines and, of course, the station master in top hat and tails. Apprentices for the GWR had to go to Swindon to be approved. It was like joining a good regiment. The Railways were privately owned and the morale of the staff was high.
>
> (Sir John Stokes, quoted in Bagwell, 1996: 133)

Nearly forty years before this speech Harold Macmillan, a former director of the GWR (and Conservative Prime Minister from 1957–63), wrote a letter about the plans to restructure the British Transport Commission, the BR holding company, suggesting:

> Why not restore the old names and titles? For instance, Western Region should be called the Great Western Railway. The head of it should be called the General Manager, as he always was. It would also give great pleasure if the old colours were restored. Our men used to be proud of their chocolate brown suits and all the rest; ...

> The regimental system is a great one with the British and it is always a mistake to destroy tradition. I am quite sure from my own talks with old friends in the G.W.R. that they would welcome recovering their identity. They don't care about who owns the shares, what they care about is their own individuality.
>
> (Harold Macmillan, quoted in Gourvish, 1986: 571)

In such discourse, the past is used to conjure up an image that will resonate with popular memory, playing on a nostalgic hankering for an indeterminate time in the country's history when it was more sure of its identity. The railway here plays an important part in framing such a view, in particular because of its association with a lost rurality (Payton, 1997; Sykes *et al.*, 1997). Davis (1979) has argued that nostalgia tells us more about the present than about the past, and that the emotion is stimulated by a discomfort with the present. The past under such an optic is more real and whole, while at the same time simpler and more intelligible. The political subtext to this sort of imagery is the desire to return to a situation when the workers were loyal to their company and, in management's view, had not been led astray by the lure of trade union promises. Here, occupational identity is seen one-dimensionally, rather than the complex differentiated reality that we have seen in previous chapters. The quotes from Stokes and from Macmillan, and indeed Major's views, demonstrate the way in which the past is malleable within popular, or in this case political, memory. Here is the overt celebration of the very thing that was central to the neoliberal critique of BR and the wider public sector: the nostalgic hanging on to the past. Nationalisation then becomes an interregnum, a foreign, almost un-English, import that should be exorcised in order to return the nation to its correct historical trajectory. Perhaps the difference between the nostalgia noted in Chapter 3 and that here is that for the Conservative politicians engaged in privatising the railways, there was the desire to actively return to the past rather than simply reflecting on it while mourning its passing.

Whilst there may have been a rhetorical sentimentalism around the possible form of privatisation, the mechanism by which the change in attitudes would occur was seen as being fundamentally market-driven in nature. The privatisation process led to the creation of a separate track authority, called Railtrack, and the division of the three passenger sectors (Intercity, Network Southeast and Provincial) into 25 Train Operating Units (TOUs). In addition, the freight sector was split into several parts, with eventually over a hundred new companies created to take over the functions of BR (Harris and Godward, 1997; Freeman and Shaw, 2000). Crucially the architects of privatisation deliberately designed both vertical *and* horizontal fragmentation in their plan, with markets and contracts replacing hierarchical command and control strategies that had been a feature of railway management across the world since the beginning of the industry in the 1820s. It is worth reiterating that fragmentation was seen as a fundamental part of changing the culture of the industry, not as some unforeseen or unintended consequence of the process.

As part of the separation process in the formation of Railtrack, 641 contracts were created between the new company and its suppliers and customers (Edmonds, 2000: 59). An illustration of the amount of paperwork is shown by Ivor Warburton, director of the then Intercity West Coast TOU, in an interview for an edition of BBC television's *Panorama* on rail privatisation (screened on 12 December 1994):

> I have 4,500 staff as my main asset, desks, computers and other office equipment and a 22-foot high pile of paper that's growing daily as we move towards completing the 840 contracts that complete the train company of today.

In the same programme Chris Green, the then director of Scot Rail, noted his 360 contracts, with each individual station having its own separate one, 100 pages in length. There were 12 legal and consultancy firms engaged during the process of privatisation and Bagwell (1996: 148) estimates that the combined cost of legal, financial and public relations fees was £48 million.

Many within BR were unhappy with this structure for, it was argued by senior management, if privatisation were to go ahead, the vertically integrated sectors themselves should be sold off intact. As John Edmonds, chief executive of Railtrack between 1993 and 1997, reflects:

> by the late 1980s, under Bob Reid's leadership, there remained only the core railway activities within vertically integrated businesses. With privatisation by then on the agenda, it was generally assumed within the industry that these businesses would form the basis for the privately-owned companies although the resurrection of the old regionally based companies was advocated by a few romantics.

> (Edmonds, 2000: 57)

Crucially, senior BR managers were not all necessarily opposed to privatisation in itself. Rather, they objected to the fragmentation of the industry and the loss of vertical integration between constituent parts which, it was believed, would have serious operational and safety consequences. Much of this critical internal comment was not publicly aired as it would have jeopardised future careers. However, in the *Panorama* programme on rail privatisation referred to above, it was reported that in a meeting between ministers, their civil servants and the top 45 BR managers, there was unanimous opposition to the plans from the professional railway personnel. The programme cited an inside source as saying:

> One of them [*a BR manager present*] said they had been cycling over a tight rope over Niagara Falls for the past four years, now they were being invited to do the same thing only backwards.
>
> (*Panorama*, 12 December 1994)

The overwhelming impression given by the managerial and supervisory grades I interviewed for my research was that they were anti-privatisation. During 1993 the Transport Salaried Staffs' Association (TSSA) carried out a survey among their BR executive grades members, which found that 91 per cent of those replying were not in favour of privatisation (Strangleman, 1998: 282).

In spite of, or possibly because of, this opposition, the Government ploughed on with its more radical plans. John MacGregor for instance, one of the ministers responsible for enacting the legislation, was still unhappy with the BR record even after a decade of change which had included the creation of the 'market responsive sectors'. Speaking in the Commons in 1993 he stressed:

> [BR] combined the classic shortcomings of the traditional nationalised industry. It is an entrenched monopoly. That means too little responsiveness to customer needs ... Inevitably also it has the culture of a nationalised industry; a heavily bureaucratised structure ... an instinctive tendency to ask for more taxpayers' subsidy.
>
> (MacGregor, quoted in Bagwell, 1996: 139)

MacGregor (Department of Transport, 1994) later wrote of the need to improve the quality of services by moving them into the private sector; by this time it was taken as axiomatic that nationalisation had failed. The aim was to allow management to 'exercise its entrepreneurial talent ... loosening Government control [and] injecting competition and introducing market forces'. MacGregor ironically went on to praise what had been achieved under public-sector management:

> Privatisation is no criticism of British Rail or its past achievements, which have been considerable.

(Department of Transport, 1994: 1)

And in 1995 Brian Mawhinney, the third of four ministers involved in the privatisation process, was also critical, arguing that even after 14 years of greater customer focus:

> the old arrangements were flawed because they were producer-led ... It was the same old nationalised industry story: the management taking all the key decisions on a centralised, bureaucratic basis and without a stable financial regime. The command economy with a vengeance. This is not a criticism of BR management. It is a criticism of the structure.

(Department of Transport, 1995: 4)

Government ministers believed, therefore, that competition between the fragmented parts of the industry had to be ensured. Ministers were keen not to see a repetition of earlier privatisations, in which monopoly service providers such as British Telecom and the water companies were privatised with their market dominance intact – a major source of criticism from opponents of the policy. The final structure of rail privatisation was a compromise between on the one hand, free marketeers and Government ministers who favoured maximum competition in the industry, and on the other hand, civil servants and those in the private sector. The latter, wishing to bid for parts of BR, were alarmed by the uncertainty that unlimited competition represented (see Freeman and Shaw, 2000). Here we have the irony of politicians and libertarian academics arguing for the construction of a perfect market, with the potential entrants and players in that market threatening to refuse to take part because of the lack of predictability. In the end the Government compromised in various ways to ensure that the Treasury obtained a reasonable return on BR assets, while 'irreversibility' was achieved by making it almost impossible for any new administration to return the industry to State control. 'New Labour' were reluctant to renationalise other former industries, but in the 1995 Railtrack flotation prospectus they claimed that on their return to office they would seek to establish a 'publicly owned, publicly accountable railway'. What made this unlikely in practice was a combination of the thousands of legally binding contracts, the fragmented nature of the new structure and the sheer cost of the exercise (see Wolmar and Ford, 2000).

Thus in the preparation for privatisation we see a subtle shift in the interpretation of the past at the political level, with the blending of positive and negative views of history. State ownership of the industry was still cast as having failed, but now there was the possibility of freeing the entrepreneurial talent that had lain fallow since the days of the private companies. Nationalisation (in this reading) could never have succeeded because it was outside the market system. We turn next to a consideration of the way several units of the BR organisation made sense of their inheritance during the process of privatisation. We do this by examining separately the themes of corporate culture and image.

Culture change

In the lead-up to, and following on from, privatisation, the 'past' has been dealt with by a continuation of this blurring between positive and negative, nostalgic and nostophobic. One of the main foci of criticism regarding negative aspects of the past has been the issue of corporate culture, with older workers singled out as being unwilling to change, and one of the key factors in the supposed failures of State control. Such a concern with culture reflects wider debates on the subject in the fields of management and organisational theory (Deal and Kennedy, 1988; du Gay and Salaman, 1992; Willmott, 1993). Munro (1998: 219), for example, talks of 'culture as a drag', with social networks, norms, values and traditions being seen as unwelcome contingencies by management. As has been argued elsewhere in this book, the interest in culture change can be interpreted as the desire on the part of management to create a blank slate, where previous contingency is abolished and organisations, work patterns and therefore culture can be made anew. Whilst BR itself had attempted culture change programmes in the public sector, as had LU, privatisation gave new momentum to this process. The passage to the private sector was seen, by both politicians and managers, as the chance to address the remaining barriers to flexibility. In a speech reported in the railway press John Welsby, the last Chairman of the BRB, said:

> One of the objectives of privatisation – rarely openly declared but nonetheless real – was to loosen up some of the rigidities of the

industry's working practices. 'Breaking the power of the unions' is certainly how some people would put it. I would say it was the opportunity to bring working arrangements more into line with the norms applying at the end of the twentieth century.

(*Rail*, 1998, 324: 38)

Though not the only reason for privatisation, as Welsby indicates, there can be little doubt that one of the main advantages that was projected from the move of the industry into the private sector was the marginalisation of organised labour through the breaking down of collective structures of industrial relations. A theme that constantly emerged in public pronouncements on 'railway culture', as well as in interviews carried out with managers in the industry, was the issue of the age profile of workers, and the perceived conservatism of older workers towards accepting more flexible working arrangements. A pamphlet on rail privatisation from the influential Institute of Economic Affairs (IEA) stressed how liberalisation might help speed culture change:

The fact that 70% of British Rail drivers will retire by the year 2000 may appear to be unfortunate because of new training costs, but it means that changes of attitude can occur through the natural process of replacement. For instance, if new, young staff are recruited by independent companies, it is not certain that they will choose to join the established rail unions, or any union at all.

(Glaister and Travers, 1993: 40)

The implication is clearly that older workers are less pliable than younger ones, partly because of their connection with trade unionism. In addition to this demographic 'time bomb', there was the 'happy' coincidence of the contraction of the industry, which was part of the very process of privatisation, with most if not all freight and passenger units shedding staff. Over 40,000 were made redundant from BR between 1991 and 1996 (*Guardian*, 3 August 1998). In Loadhaul, the chance to reduce levels of older staff was eagerly taken up.[2] As one Loadhaul human resource manager I interviewed in 1995 explained:

Ten years ago when I first joined we were having a mega panic because of age profiles of drivers. They were all nearly in their sixties … but because of concepts like Train Crew Agreement, rationalising of staff, reduction in staffing, Driver Only Operation and Variable

Rostering and all the other things that have reduced staff numbers, we are at a situation currently where age profile isn't too bad. A lot of the older people have gone under voluntary severance schemes and a lot of those still here are waiting to go quickly because we have got an attractive severance package at the moment which they can see won't last after privatisation. So they are keen to take advantage of that ... we are rapidly getting to the stage where we have got a young work force.

In this sense the older workers are 'softened up' before being offered 'voluntary redundancy', a recurring practice in British industry over the last two decades (see Roberts, 1993; Turnbull and Wass, 1995). Another Loadhaul manager I interviewed in 1995 spoke of the problems that age presented:

The young people don't perceive as much as the old people ... A lot of the generation of people we have got were from that generation [*steam days, pre-1968*], people older than 55, they are the real old stalwarts, they are the people who create the biggest barriers – not maliciously, it's just there.

As Munro (1998: 224) has said:

Loyalty, and the pride that rests with tradition, is reduced to the status of an *attitude* ... a barrier to progress.

Similarly, a Loadhaul industrial relations manager (interviewed in 1995) differentiated between older and younger workers in the driving grade:

Some of them realise they can't hang on to too much tradition, the old ways would not be helpful to themselves or to the public ... but it doesn't happen overnight. You can't destroy 30 years. Most of our employees now are under the 35 age group ... we've still got some people well over 45, it's the older workforce [who are the problem].

Shortly after these interviews were carried out, Loadhaul, along with the majority of the former freight sector of BR, was purchased by the US-owned English, Welsh and Scottish Railways (EWS). EWS have made even more explicit moves to 'change the culture'. Ian Braybrook, the former head of Loadhaul, was made managing director of the new company, and here summarises his thoughts on culture:

> As for staff-management relations, it is a fresh culture we need to acquire. The lack of the 'them and us' mentality, getting people to believe in the company, to want to work for the company ... There is an old business school dictum which says: 'To turn a company that is performing badly financially takes three years; to turn round a company which has inadequate systems and IT takes six years; to change a culture takes 20'. Well we haven't got 20 years. We aim to change the culture in three or four.
>
> (*The Railway Magazine*, August 1997: 49–50)

In other parts of the 'new' railway, the process of redundancy was similarly coupled with a targeted attack on the 'older culture'. Within Railtrack a corporate culture change programme was enacted, again with explicit criticism being made of the past:

> a company-wide improvement programme called C. Change to deliver the culture change that will make us a more flexible and customer focused organisation.
>
> (Railtrack, 1997: 3)

There are many other examples of companies and groups within the 'new railway' adopting culture change programmes – GNER, Virgin and Northern Spirit to name but three (see Strangleman, 1998: 239–95). Thus the focus on culture within the commercialisation and subsequent privatisation process can be interpreted as being informed by a nostophobic sense of the past. In such a view the responsibility for past failure is laid directly on the agency of those who have worked in the industry, be they management or workers. This group is interpreted as living in the past, holding nostalgically to the old ways, to tradition, while the corporate world has moved on around them. It therefore falls to the nostophobic to save the industry from its own nostalgia.

Image change

A second area in which privatisation has seen the new companies distancing themselves from the BR past is that of corporate image and identity. In many ways this was again a continuation of the process that had begun during the 1980s, with a proliferation of new liveries and logos, perhaps best illustrated by the Intercity sector, which actively sought to distance itself from the rest of the corporate body (Vincent and Green, 1994). With privatisation underway, the various subsidiaries of BR went even further. Loadhaul adopted a new identity, and company literature from 1994 stressed it was keen to:

> develop an identity distinct from British Rail and from the other former Trainload Freight businesses who will soon be our competitors.

This identity was to last for less than two years, as Loadhaul then became part of EWS.

Perhaps a more ambiguous example of corporate image change was in the marketing strategy of Intercity East Coast, the unit of BR which ran trains on the east coast main line from Kings Cross. Interestingly it was to the past that Intercity East Coast, and its brand consultants Interbrand, looked:

> Interbrand reviewed the company's previous visual identities, selecting an early LNER concept as being particularly interesting. Interbrand's aim was to develop an elegant signature and symbol, combining modernity with tradition, which would clearly communicate that the service offered by East Coast is premium-quality, efficient and friendly.
>
> (*Livewire*, August–September 1995: 7)

This image did not last long. In March 1996 the east coast franchise was awarded to Great Northern Railways Limited (GNR), a subsidiary of Sea Containers. In an interview Christopher Garnett, chief executive of the new company, discussed the choice of the name GNR and was asked whether he was aware of that company's past:

> it's tied-in absolutely to the brand ... Yes, we'll build on the past, but I think we have to have colours and an image that are relevant to the 1990s. The product has to be the fastest trains in Britain – the East Coast has always led the way there and we will continue to do that ... We'll be a trail blazer – but we're not going back to the past – that's why we haven't used the name London & North Eastern Railway.
>
> (*Rail Magazine*, May 1996, 278: 25–6; my emphasis)

Note the contradiction here in Garnett's rejection of a 'return to the past', whilst leading a company consciously named after one founded in the mid-nineteenth century. History is being used as a 'brand' attached to a 'product', something that is itself capable of being marketed. As Jacques Le Goff (1992: 95) put it, 'memory has thus become a best-seller in a consumer society'. GNR was forced to change its name to GNER (Great North Eastern Railway) for legal reasons in 1996, and later that year it announced the appointment of a US design consultancy. In a 1997 issue of *Livewire*, Garnett proclaimed:

> The most visible change is of course the new name – Great North Eastern Railway – and colour schemes to match. Inevitably, this will be shortened to the initials GNER, and I expect (and hope) that this acronym will be as well known and respected as the LNER was some 50 years ago ... New names and colour schemes make the most obvious statement about moving away from the British Rail past.
>
> (*Livewire*, December–January 1997: 6)

In this later discussion of history, the LNER seems to have found favour. It is interesting to reflect that these pronouncements were made at a time when the last of the former LNER staff would have been retiring from railway service – those that had not already been culled in corporate culture change programmes. It was during this period, in 1996, that a GNER leaflet was issued showing a 'new' liveried train symbolically emerging out of an 'old' Intercity East Coast one under the banner 'The new golden age of rail travel is arriving' (see Figure 5.1). Later literature from the same company continues this theme:

> We are proud inheritors of traditions of speed, safety and service going back nearly 150 years. Now under private management, we intend to build upon these traditions and strive for new levels of excellence in the age of high speed train travel.
>
> (GNER Passenger's Charter leaflet, 1997).

Tradition was re-emerging as a positive quality which the railway had apparently long been noted for. The 1996 GNER leaflet mentioned above (Figure 5.1) reads:

> Great North Eastern Railway ... is a new name with a sense of the history, the romance, of the great days of railways. Our aim is

The new golden age of rail travel is arriving

Figure 5.1   'The new golden age of rail travel is arriving'. Promotional leaflet produced by GNER in 1996. (GNER)

to recreate the golden era of railway travel by setting the highest standards of service, speed and quality. The name comes with new train liveries and corporate design.

The new GNER identity (see Figure 5.2) made use of cast metal plates which mimicked older coats of arms that former railway companies adopted. As Hobsbawm and Ranger (1992: 1) have argued in their discussion of those who seek to 'invent tradition': 'where possible, they normally attempt to establish continuity with a suitable historic past'. This theme of evoking romance was also highlighted in a 1997 edition of GNER's passenger magazine *Livewire*, in an article about an advertising promotion:

> The campaign builds on the golden era of train travel, and the quality of service evoked by the GNER name ... GNER also recognises that rail travel can be as much an emotional experience as a practical one.
>
> (*Livewire*, December–January 1997: 8; my emphasis)

Thus, we have the direct annexation of emotions, particularly romantic and nostalgic ones. What is directly criticised as a negative trait in the 'old culture' is now overtly traded on in the 'new market-orientated' era. Other companies harnessed these romantic associations of railways, but they were mixed with a negative reading of the past. The Virgin Rail group made much of its brand attributes, supposedly reflecting younger, fresher values, and began re-imaging its fleet of trains in its house colours. After some initial consensual references to combining the best of BR with the skills of Branson's group, Virgin has been forced to reappraise its relationship to the past. This reflected the

Figure 5.2   GNER coat of arms, 1996. (GNER)

enormous amount of criticism levelled against the company's performance on its west coast mainline franchise (see Wolmar, 1997). Richard
Branson, writing in *Hotline*, the Virgin Trains passenger magazine, in
1997 spoke of:

> facing the daily challenge of undoing the 30 years of neglect,
> demoralisation and structural decline we inherited.
>
> (*Hotline*, Autumn/Winter 1997: 2)

Ironically, later on the same company, like GNER, also looked to the
past more positively. In the Spring 1998 issue of *Hotline*, a poster campaign is described as:

> Designed in a 1930s retro style ... the three posters set out to rein-
> force the Virgin Trains brand image ... We blended the classic
> romance of railway travel, as embodied by the 1930s, with a modern
> take on travelling by rail, all in Virgin's distinctive red and black
> colours.

Other companies have been quick to find value in their past, with
Great Western Trains and Great Eastern Railway being the best examples. Railtrack, a company which was one of the earliest to attempt
to enact a 'culture change', has sought to engage in a celebration of the
nostalgia associated with the railways. In its statutory accounts for the
financial year 1996–7, Railtrack (1997) commissioned a series of posters
(see for example Figures 6.1 and 6.2) from a number of artists who borrowed directly both stylistically and for their subject matter from past
examples of railway poster art, most notably the work of Terence
Cuneo (see Figures 4.2 and 4.3, and also Harris, 1997). In its next set of
accounts (Railtrack, 1998), a caption on a similar poster boasts of
'Reviving a legend' in connection with the company's repainting
of the Forth bridge (see Figure 5.3).

The use of image in the rail industry around the period of privatisation displays a subtle reworking of the past, with *heritage* and *tradition* emerging as commodities that can be packaged and sold. The
immediate past was still kept in the background as a reminder of
supposed previous failure, an era that could be conveniently resurrected at moments when the present seemed less than perfect. The
new players in the industry simply alluded to '30 years of neglect' or
'a legacy of mismanagement' as justification for their own poor
performance.

Figure 5.3 Reviving a Legend. Railtrack promotional poster by Andrew Davidson and David Partner, reproduced in Railtrack Annual Report and Accounts, 1997–8. (Railtrack plc)

Conclusion

By 1 May 1997, John Major had achieved his wish to make railway privatisation irreversible. The whole of BR had been sold off to the private sector. Railway privatisation was the most complicated of all those undertaken by either the Major or Thatcher administrations. In some cases individual units – such as the freight and the track maintenance companies – were sold to the private sector. Railtrack, the monopoly owner of railway infrastructure (track, signals, stations and bridges), was floated on the Stock Exchange as many of the other utilities had been. Finally, the 25 Train Operating Companies (TOCs) were subject to franchise. The private sector and management buyout teams bid to run the service for between five and 15 years on the basis of their requirement for subsidy. Unlike earlier sales, however, the rail industry continued to require massive amounts of public subsidy to operate loss-making routes and services. While the TOCs were the direct recipients of State aid, all of the other parts of the industry were reliant on this subsidy either directly or indirectly. The TOCs rented their rolling stock, and paid access charges to operate on Railtrack's system. In turn Railtrack used this money to pay other privatised parts of the former BR to maintain and renew its system (i.e. track and signalling). Unlike BR, however, the private companies owning or running services had guaranteed income streams from the public purse, which gave them a degree of stability and predictability that the State-owned company had never enjoyed.

Despite this complexity, Government ministers believed that the system created a more dynamic environment for would-be entrepreneurs to flourish in. The dead hand of monopoly in the guise of BR had been removed, and with it the bureaucratic managerial culture. At the same time a direct aim was to change the workplace culture in the industry – seen as one of the last bastions of restrictive practices and union militancy. By fragmenting the industry vertically and horizontally into so many parts, it was assumed that labour collectively and individually would become more 'flexible' and 'realistic' in their demands. Managers of these new companies and units would now have the discipline of market forces, unlike their BR predecessors who could always go back to the Government for help. In the next two chapters we look at the experience of labour during the process of privatisation and its aftermath.

6

A Brave New World: Working for a Privatised Railway, 1979–2001

> The poison called History perpetuates misunderstandings and suspicions which would otherwise disappear under the present new conditions ... Happy industries seem to be those with no history just as happy nations are.
>
> (Ferdynand Zweig, *The British Worker*, 1952: 86–7)

Introduction

The organisational theorist Martin Parker (2000) noted the 'explosion of enthusiasm' since the 1980s for writing on organisational culture. Management writers have encouraged their readers to think of themselves as organisational anthropologists able to tackle 'soft' people issues, rather than concentrating on company structure. Excellent, successful and dynamic companies are those with the 'right culture'. Nowhere has this general shift in focus been more warmly embraced than in the UK railway industry (Bate, 1990, 1995; Guest *et al.*, 1993). 'Cultural' explanations for failure have been attractive as they conveniently neglected the facts of long-term underinvestment in the industry, putting the blame instead on the norms and values of railway workers themselves. If the railways were to succeed, they required a culture change, one where values of entrepreneurship and commercial sense displaced the public-service ethos. As we saw in Chapter 5, 'culture', and the need to change it, structured the thinking of successive politicians involved in the privatisation process, and of the new managers who now run parts of the former State-owned British Rail. But how do we understand culture change programmes? How did workers themselves experience this change? In what ways has the fragmentation of the industry affected the workforce? Finally, what are the contradictions in this desire to 'change the culture'? We begin this chapter by trying to understand the impulses behind the focus on culture.

Understanding culture change

At its simplest, the desire to 'change culture' is predicated on perceived organisational failure. Usually a company's personnel is blamed, and while sometimes this can include management, more often it is the workforce that is the target. This way of viewing an organisation is based on a peculiarly narrow conceptualisation of past, present and future. History, however venerable, is judged as no longer providing an adequate guide for the contemporary world. It follows that those who seek to defend the past or value the insights it might offer are accused of living in the past, of being nostalgic, sentimental and ultimately irrational. If the organisation is deemed to have failed, those who defend it are themselves an obstacle to progress – where progress is understood narrowly as something *different* from the past. At the heart of this critique is the belief in a crisis of Western capital, and in particular the way in which the workforces of the older industrialised countries became lazy and unproductive in the era of the long postwar boom (Nichols, 1986). The desire of management and politicians is to realise the goal of flexibility and to create an entrepreneurial culture amongst the workforce, something believed to have been a universal feature of Asian economies, before the crisis of the late 1990s (du Gay and Salaman, 1992; du Gay, 1996).

The problem for management is that it is trying to enact change in an existing workplace in the context of established norms, values and patterns of work. Traditional industrial sociology has understood this tension through the conceptualisation of the employment relationship, most notably Goodrich's (1975) idea of the 'frontier of control', or Edwards' (1979) theorisation of the 'contested terrain'. Both of these writers attempted to model the boundaries between capital and labour at the point of production, the workplace. Rather than a straight exercise in power, as perhaps only the crudest of Marxists would argue, the employment relationship takes the form of a *negotiated order*, in which unequal power relations are continually mediated and contested over time. Thus power and control over work are not simply determined by managers or supervisors but are understood as contingent on, and mediated by, a range of factors – labour market conditions, skill, technology and even the personalities of workers and supervisors (Brown,

1988, 1993; Roberts, 1993). The limited power that labour can exercise at the point of production acts as a resource for resistance to the potential that management has to organise and reorganise the workplace as it wishes. Worker control over the labour process is also a product of individual and collective knowledge about work. In this sense employees are embedded in their work, and their authority over it is partly legitimated by their very attachment to it.

Culture change programmes are, then, the attempt to tackle the problem of an established workforce having the power to resist or block change. They do this by pushing back the frontier of control between capital and labour on the level of the shop floor, or the organisation more widely. Management are attempting to change the architecture of the employment relationship, to replicate, in the context of an *established* site of work, the seemly attractive traits of a workforce in a *new* workplace or greenfield site. Culture change therefore becomes the process of the *greening* of a brownfield site. The greenfield site enables management to select and determine the workforce it wants, giving it power to exclude those with the wrong attitudes or traits (Garrahan and Stewart, 1992; Graham, 1995). This wish for a 'green' workforce is reflected elsewhere in sociological literature on work. For example, at several points in du Gay's (1996) discussion of the contemporary retail sector, he explores the way in which some of the managers he interviews dream aloud of a corporate tabula rasa, a blank slate on which new organisational identities could be written. Here, du Gay (1996: 150) describes the thoughts of a company director:

> Ideally, he would like to have seen all current sales staff removed from their positions and replaced by completely new recruits better able to fit the company's desired 'culture'.

Catherine Casey (1995) also writes about the 'designing' of corporate culture, and the desire to produce an organisation full of 'designer employees'. In seeking culture change, managers are engaged in the search for their ideal employee, a designer worker. This is a worker who embodies entrepreneurial flair, is flexible, adaptable and committed, while at the same time is shorn of the negative aspects of work culture. Often, as we saw above, culture change programmes specifically target older members of the workforce as embodying these latter traits, and such a focus has important and interesting implications for this chapter. Older workers are perceived as being a major barrier to corporate success, wedded as they are to 'the old ways', as one Loadhaul industrial relations manager I interviewed in 1995 put it.

The social anthropologist Margaret Mead (1978) examined generational relationships within and outside the family in the context of North America during the 1970s. She was particularly interested in what happened when traditional forms of socialisation – what she labelled postfigurative cultures, where the past is the guide for present and future behaviour – were disrupted. Mead (1978: 49) noted the way nuclear families were highly flexible, possibly because they carried less in the way of intellectual baggage about the past, and she draws the parallel with corporations:

> In large organizations that must change, and change quickly, retirement is a social expression of the same need for flexibility. The removal of senior officers and elderly personnel, all those who in their persons, their memories, and their entrenched relationships to their juniors, reinforce obsolescent styles, is parallel to the removal of grandparents from the family circle.

She linked this discussion with the absence of grandparents in the context of immigration to the USA:

> the child's experience of its future is shortened by a generation and its links to the past are weakened. The essential mark of the postfigurative culture – the reversal in an individual's relationship to his child or his relationship to his parents – disappears. The past, once represented by living people, becomes shadowy, easier to abandon and to falsify in retrospect.
>
> (Mead, 1978: 49)

Culture change programmes, especially where they are associated with redundancy packages, often act to remove older workers from the new order under construction. These attitudes and assumptions about age and generation tell us as much about managerial views of younger organisational members as they do about their perceptions of older ones. While there might be a rhetorical claim that younger workers are more dynamic and entrepreneurial, it seems that what is really valued amongst this group is docility or passivity in the face of management demands and dictates. The logic of such an approach is that what is desired in the ideal new worker is a kind of adolescence – willingness, enthusiasm and commitment without experience.

Richard Sennett (1996), in his discussion of the problems of urban life, explores the tensions between notions of adolescence and adulthood. He is critical of the way urban planners try to engineer, or over-prescribe the use made of particular parts of the city, likening the purity of residential communities, especially the suburbs, to the exclusionary identity often adopted by youth groups during adolescence. This adolescent identity is one where homogeneity is prized and difference is viewed as a threat. Sennett argues that in seeking to exclude the impure, the complex and the diverse we lock ourselves into a fragile and less complete relationship to our surroundings, one that is constantly under threat from dilution. The fully adult condition is one in which experience challenges, and ultimately undermines, the attempt to hold onto a pure, fixed notion of identity, and complexity and diversity are celebrated. Culture change programmes would seem therefore to be a similar attempt to engineer adolescence in the workforce, in that they seek to exclude experience in the collective workforce as well as in individual members of it, and are thus able to deny the legitimacy of alternative readings of history founded in knowledge of the past.

Like Sennett, Peter Marris is interested in the process of historical experience, and argues that the fully adult identity is one that is necessarily experienced, and is also conservative with a small 'c':

> We grow up as adaptable beings, able to handle a wide variety of circumstances, only because our sense of the meaning of life becomes more certainly consolidated ... it is slow, painful and difficult for an adult to reconstruct a radically different way of seeing life ... in this sense we are all profoundly conservative, and feel immediately threatened if our basic assumptions and emotional attachments are challenged. The threat is real, for these attachments are the principles of regularity on which our ability to predict our own behaviour and the behaviour of others depends.
>
> (Marris, 1974: 9–10)

Assimilation of knowledge about the world is by its very nature conservative, as is the process of rendering the unfamiliar familiar. There is equally an impulse to preserve predictability. In culture change programmes, management are effectively attacking what it is to be adult, denying the legitimacy of experience. To extend the family metaphor, management take on the role of parent to the worker's adolescent or even childlike identity. Culture change initiatives are essentially unitarist in their aim, as they abolish dissident interpretations of the past either by the direct removal of critical voices, or by attempting to silence those who remain. They are, as Grint (1995) noted, a kind of domination over the silenced. We will return to some of these ideas later on in this chapter.

Experiencing culture change

We have established that railway management talked extensively about culture change during the 1990s. Now we need to ask to what extent this was simply a rhetorical exercise, or whether there has been a determined effort to shape cultural practice. There can be no doubt that privatisation was preceded by a mass exodus of senior staff from the industry. It is difficult to be precise about the number of jobs lost in the sector during the 1990s, partly because of the fragmentation of the industry itself. Estimates vary, with some claiming that between 1991 and 1996 40,000 redundancies occurred within BR or its successors, at a cost of £610 million to the public purse (*Guardian*, 3 August 1998). Since 1996 still more jobs have been lost as the hundred or so organisations set up as part of the privatisation process have reorganised, rationalised and consolidated their operations. Ian Jack (2001: 59) has suggested that between 1992 and 1997 the number of workers employed on Britain's railways fell by 67,000, from 159,000 to 92,000. This collapse in numbers was the result of rationalisation of work, lost contracts and competitive tendering exercises. On London Underground, a quarter of all staff – over 4,000 workers – were made redundant as part of the Company Plan culture change exercise of the early 1990s. These redundancies in the industry were overwhelmingly achieved 'voluntarily' and in large part took place among older workers at whom redundancy or early retirement packages were especially targeted, based as they were on seniority. While these 'carrots' may have been attractive, some fairly heavy sticks were also involved in ensuring sufficient numbers of workers left the railways. Many of those I interviewed in the mid-1990s spoke directly of their own feelings about change, or of those of older colleagues who had chosen to leave:

> Most of them really really wanted early retirement, just to get out; just sick of the way the railways were being run. In the '80s and '90s the driver wasn't treat with any respect by management, you were just a driver and 'You get on with it and you do it' sort of thing. An awful lot of young managers who hadn't done sort of

footplate service were getting managerial posts. And so they had a bloke telling them what to do when they had never done it themselves.

(John Oliver, Loadhaul driver, interviewed 1995)

Many workers discussed wanting to leave the industry during or after privatisation. A Railtrack signalman on the point of retirement said:

There's me and another old signalman about to retire. He can't get away quick enough, and my two months is going to be the longest two months!

(John Porter, Railtrack signalman, interviewed 1995)

Why is it that older workers present such a problem to management? The key is that older workers in established workforces embody a moral order that is legitimised by their seniority and experience 'on the job'. Attributes such as knowledge of the labour process, agreements and procedures, coupled with respect from other colleagues, combine to form a worker embedded in a work role and able to draw on considerable resources. This authority that older workers could exercise was nicely illustrated during my fieldwork in 1995 when I shadowed a Loadhaul train crew leader (TCL) for a day. While waiting for him in a corridor at a train crew depot, I witnessed an argument between a driver in his late-forties and a young train crew supervisor (TCS). The TCS had allocated to a junior man a job that, according to local agreements, should have been this particular driver's. It was clear from the interaction that the supervisor – knowing he was in the wrong – was very wary of the driver. The respect that supervisors showed to older workers was also manifest in a comment from a Loadhaul instructor:

The one thing I was told when I first came to the inspectorate side was by one of the old boys who's just finished: 'The best way for you to get on at the moment is to ask the old hands, tell the young hands'. I adopted that policy. It's amazing how once you start demanding [something], they dig their heels in, the job stops.

(Loadhaul instructor, interviewed 1995)

All the driving staff I talked to noted the difference in the way the older drivers had been treated and addressed. In their absence, a change in the attitude of management and supervisors was commented on:

> It's totally different now, it's a younger job. Most of the older hands have retired through voluntary redundancy, and coming of age at 65, I mean the age profile now, the average when I started was mid-fifties. But now I would say down to mid-twenties. It's a young man's railway but then again we're a smaller workforce. It's good and bad. In one way you're working with people of your own age, in another way there are serious erosions of the job – we're more in line with lorry drivers!
>
> (Gary Hill, Loadhaul driver, interviewed 1994)

This theme of the status and respect which former workers had enjoyed, compared with current conditions, was one that constantly re-emerged in discussions. As one driver in his thirties put it:

> These days management just say 'If you don't want to do it pick up your cards on the way out'. They could never get away with it with the older drivers.

I asked a group of younger drivers in their mid-twenties whether, if the older workers had still been on the job, they would have accepted the changes in working conditions and supervision. One replied:

> They never would've accepted it, never! And management wouldn't have pushed it either.

Almost irrespective of the validity of such claims, the younger workers perceived the older members of the workforce as having resisted management pressure, and that as a result management at times had to respect that. Furthermore, such resistance was in part predicated on the legal requirements within the industry that certain grades were not only conscious of, but had historically helped to create and police. It was almost as if authority resided in the older members of the workforce and acted as a powerful resource in and of itself. The presence of some workers, and the sense of awe they inspired, was touched on in Chapter 3, with the young station master Ferneyhough (1983: 87) describing 'Engine drivers [as] being formidable fellows ...'.

I found in my fieldwork that this contrast between the treatment of younger and older workers was not restricted to the former BR companies, but was a theme touched on by many LU workers too. Huw, a worker in his forties, highlighted the changed managerial attitude to younger workers:

> They want to get rid of the old school that knew the job. Most of the people of my age group and above, and younger ones want to get out. It's going back to the Victorian days. The managers just laugh at you now if you say you're going to get a union rep: 'Do what you like'. It was like a family and people used to stick together; it broke down completely. By getting rid of all the older guys and having a lot of the younger guys who don't know any different, now they have a young workforce that weren't brought up in the railway proper like I was. And they [*management*] bully them and they get away with it. I've seen a lot of young blokes being bullied 'cos they are too frightened to say anything. A lot of staff are really frightened of the management, and the way they are treated.
>
> (Huw Evans, LUL signal operator, interviewed 1995)

This same worker drew parallels with the working conditions enjoyed by fast food restaurant workers, and indeed a number of interviewees noted that McDonalds had been used as an example that senior management wanted to follow (Strangleman, 1998). This linkage between flexibility and youth was echoed in an interview with an RMT official in the context of the now privatised national rail system:

> They [*the train companies*] have got a policy of recruiting as much as possible from the street, you know, young keen people that they think will be completely malleable, won't be trade-union minded and they will be able to do what they want.
>
> (RMT official, interviewed 1998)

There would seem to be evidence to suggest that there has been a deliberate policy to 'green' the workforce, by both removing older workers through early retirement and voluntary redundancy schemes, and actively recruiting new younger workers where there are vacancies. But this is not to say that this process is simple or without contradiction. The main issue for management in the established workplace is that workers are embedded in the organisation in terms of their skills and knowledge about the labour process as well as cultural practice and norms. The theme of the frontier of control is important in understanding what is occurring. By removing some – though obviously not all – of the older, more established workers, management are excluding some of the most critical voices. Simultaneously they are changing the power relations at the point of production. Therefore, for those who stay, resistance to management is made more difficult, their power and resources in the organisation are undermined or restricted. The possibility of resistance based on custom and practice is marginalised. If the frontier of control is, as Brown (1993) has argued, a 'negotiated order', then changing the dynamics of the workforce, particularly on the axis of age, affects the relative power capital and labour bring to that negotiation. Shifts in the age profile can be thought of as the greening of the workplace, a kind of engineered adolescence. The restricted pluralism that exists in the 'brown' workplace, based on a more adult condition, is replaced with a far more unitary conception of the right to manage – management take on the adult role in the organisation, the worker that of the child.

As we saw in the previous chapter, the era of nationalisation was often used to symbolise failure, and as part of this, knowledge about railway work was also delegitimised in a variety of ways. In attaching negative meaning to the past in this way, the manipulation of the present is more easily justified. History can be useful in a negative sense:

> Dismissive backward glances, intended only to contrast the tarnished past product with the latest shiny device, mark the typical limit of retrospective analysis.
>
> (Ramsay, 1996: 157)

In the railways the implication was that those who looked back to earlier times were seen as nostalgic failures. It was as if established workers were the embodiment of that failure. As one signalman who was on the point of retiring when I interviewed him put it:

> The old management used to come up through the grades. Now the big bosses, if they've got seven years that's a lot. I know they've got degrees and this that and the other but a degree in history doesn't help you out with railway work. If the old men, old railwaymen, if they start to talk about the old days a glazed look comes over the eyes of management, they're not interested. It's 'Oh here we go again, the "Olden Days"'.
>
> (John Porter, Railtrack signalman, interviewed 1995)

What I found most striking when talking to railway workers during the mid-1990s was the way in which railway sense or culture was being actively rejected by the newly privatised companies. This was an experience that transcended the divisions between blue and white collar. Indeed it was in the context of the management grades that this process was perhaps most interesting. One manager spoke of the way history had almost come to an end when Railtrack was established, likening the position he now found himself in to that in Cambodia after the takeover of the Khmer Rouge:

> It's like 1994 was Year Zero. Anything that happened before that, they don't want to know.
>
> (Railtrack Manager, interviewed 1997)

The same person went on to illustrate his point by describing key meetings on operational matters where he and another colleague would be the only long-serving railway personnel present. He observed how younger but more senior colleagues' eyes would roll every time he tried to make a comment based on previous events or incidents that might have occurred two decades before:

> Because it happened 20 and 15 years ago, or before that, they find that threatening because they can't identify with what we're about ... Oh yes, that railway culture's been challenged, but at the end of the day, for as long as we intend to run trains, there'll be problems and by all the old hands being told to shut up and get out, somebody should say 'Hang on now, why didn't we listen to them?'.

This manager told me that in the end he usually kept quiet in such meetings rather than provoke a confrontation that could have major implications for his career.

This attitude towards workers and managers with railway experience was repeated elsewhere in the new industry. An official with the main white-collar railway union, the Transport Salaried Staffs' Association (TSSA), commented:

> Railway people weren't just managers, the managers were always very vulnerable because what you actually have got is people who have committed their whole lives to the industry. In the private sector people tend, if you are in the top dozen in a company, they

145

Figure 6.1 *Working in 'Step'*. Railtrack promotional poster by Andrew Davidson, reproduced in Railtrack Annual Report and Accounts, 1996–7. (Railtrack plc)

tend to move on, they don't stay. Three or four years and they are off. And they [*the incoming new management*] are finding people working with 20 or 30 years in a company, they say 'Well you can't be a very good manager because you've had 20 or 30 years with a company' – which is rubbish.

(TSSA official, interviewed 1998)

Privatisation, and the lead-up to it, drove the deliberate marginalisation of established workers, either through redundancy schemes or through a change in the way work was organised. It was precisely these workers' experience that was viewed as a barrier to the industry's progress. But culture change, whether intentional or not, has also been achieved by other means, and it is to these that we now turn our attention.

Work and fragmentation

Another way in which culture change was attempted was through the fragmentation of the industry, both horizontally and vertically. The new structure, it was hoped, would offer the best way of ensuring competition between the various sectors of the new industry. Labour flexibility was seen as a major goal in this regime, as it was predicted that in exposing the privatised units to greater competition, the industry's management would be forced to confront the unions and the workforce in order to reduce costs, as a way of winning and retaining contracts. In part this new environment was both the cause and effect of the process of greening in the industry. On the one hand, management were able to change the detail of the labour process because of the shifting make-up of the workforce, and on the other the continued deterioration of conditions for some railway workers meant that many left the industry because of continual upheaval and uncertainty. Importantly this fragmentation, and the insecurity it breeds, have not been felt evenly across the workforce, and the sense of insecurity has been experienced in a variety of different ways.

One of the major ironies of the whole privatisation process is the fact that the drivers have, at least in monetary terms, done rather well out of it. The drivers were the most powerful group among BR workers and liberalisation has buttressed their position. Rather than BR being the monopoly employer (and supplier) of labour, there are now 35 companies that have a requirement for drivers at a time when they are at a premium, as a union spokesperson observed:

> Frankly they [*Government and managers*] saw privatisation as a way of driving down wages, not improving them. Now what has actually happened to us is that the restructuring exercise has considerably boosted drivers' wages, because as I say we are a scarce commodity. You had a situation at the start where the driver's salary was about twelve thousand pounds, average of about eighteen or nineteen thousand. Now the lowest basic rate is £20,000 and average earnings are up to twenty-four or twenty-five thousand pounds. We have considerably increased the basic wage and the average earnings, because we have market strength, more than we had before.
>
> (ASLEF spokesperson, interviewed 1998)

Several factors have combined to create this driver shortage, all of which were related in some way to the sell-off process and the greater penetration of market forces. Firstly, most of the train companies before and after privatisation shed staff in order to cut costs. In the post-privatisation market, even more staff were offered attractive severance packages as part of restructuring deals, with some of the companies reducing their number of drivers by 10 per cent. These deals, for those who chose to stay, resulted in a change in shift patterns from anywhere between five and 11 hours. The aim was to improve the actual period of productive (driving) time in any one shift. However, some of these restructuring packages were driven through by managers with little or no experience of railway operation, and the result was that the overall establishment – the number of staff needed to fill the roster – was reduced too severely.[1] South West Trains (SWT), for instance, which operates commuter services from London to the South West of England, made 70 of its nearly eight hundred drivers redundant, but struggled thereafter to provide adequate service levels. Managers, lacking industry experience, had overlooked the need to allow drivers time to learn new routes and train on different rolling stock. To avoid additional fines the company was forced into recruiting new drivers, and paying those who remained substantial overtime and bonus payments. This included buying-out of some workers' holiday period, by offering them £4,000 each. SWT has regularly featured in the press over the issue, with the *Independent* of 23 March 1998 headlining a story '£50,000 a year puts railmen in new pay elite'.

The second factor in the general shortage of drivers has been the reluctance on the part of the new companies to train new drivers. This is also a consequence of the fragmentation of the industry. The average age of drivers was relatively high on BR, and as these workers retired the situation became even worse. Virgin Trains, for example, inherited a driving workforce whose average age was 50 (*Financial Times*, 20 October 1998). The average cost of training a driver is, according to union sources, approximately fifty thousand pounds, and it can take over a year for a driver to complete all aspects of qualification and become fully productive. Understandably individual train companies have been wary of incurring these sorts of costs only to see their staff lured away by other firms. The result has been a spiral, in which too little training has been done and so train drivers have been poached on a regular basis. An ASLEF spokesman described the league table of wages that was widely recognised among the grade and which the union circulated to its members. Where labour markets were tight, or where there were several train operating companies (TOCs) in close proximity, it was common for drivers to move between employers on the basis of differentials in pay. The introduction of market forces in the context of privatisation therefore strengthened an already powerful group.

For the non-driving grades, however, the experience of the newly fragmented labour market was almost universally negative. The essential difference between the driving and the non-driving grades remained the disparity in levels of skill or qualification. Most non-driving railway employees require much less in the way of formal training, and as a result it was amongst them that the new employers in the industry sought to implement cuts. The picture is one of great variety, with many companies adopting different labour practices. The way in which insecurity has been felt and coped with, therefore, varies accordingly. Since privatisation there has been an acceleration in the outsourcing of labour in non-core railway work – train cleaning and preparation being the best examples. But subcontracting has also been carried out in core activities as well. The track maintenance workforce – known as permanent way workers – was one of the first to experience a deterioration in employment terms and conditions. In the railway created by privatisation, Railtrack owned all the track and infrastructure, and the train operators paid to gain access. Railtrack, in turn, looked to contractors to maintain and renew the network and does not directly employ staff to do this work. At privatisation, 14 infrastructure and maintenance companies were created who bid competitively for Railtrack contracts. Railtrack aggressively sought to cut costs through squeezing contractors, who in turn achieved savings by reducing their workforce and further sub-contracting labour. The number of workers directly employed by these infrastructure companies fell from 31,000 in 1992 to between 15,000 and 19,000 in 1997 (Jack, 2001: 59). An RMT union official I interviewed in 1998 explained:

> There is a layering. There is the permanent staff, contracted staff that work for the main contractor, there's subcontracted staff on a short-term basis and that is all in a melting pot together, and it doesn't always work. Sites that were often worked in big relays are often now chaotic because half the people don't know what they are doing. It's the original BR staff that are in charge of the operation who are working with people who aren't railway trained.

He likened the situation on some contracts to the kind of casual labour usually associated with 'fruit picking gangs'. He also spoke of workers being employed cash in hand and cited examples of people 'off the street' working 36 hours without a break. Regulation was seen to offer little prospect of changing this situation:

> If a particular contractor proves to be particularly irresponsible in that respect, the main contractor will sack him and take on someone just as bad, or the same contractor under a different name. It goes on all the time. They can always get people to turn out.

This view was echoed in a piece in the *Guardian* (14 March 1998) which highlighted the safety implications of the deregulation of track maintenance, citing several former BR workers who now worked for private firms. One was quoted as saying:

> Privatisation has led to a vicious circle. Railtrack instructs the main contractors to do the work. They pass it on and get a cheaper job done, and it leads to unsafe practices. There is a steady deterioration.

Another noted the way in which the growth in contracts in all aspects of the industry had put greater strain on staff:

> Companies are prepared to take more risks because Railtrack bullies them. And in turn, we get bullied.

Figure 6.2 *Railtrack is Spending £4 million a Day.* Railtrack promotional poster by Debbie Cook, reproduced in Railtrack Annual Report and Accounts, 1996–7. (Railtrack plc)

The RMT union official I interviewed in 1998 commented on how this had resulted in a loss of morale and commitment on the part of the former BR workforce:

> On the Permanent Way, a maintenance gang had a particular section to look after. They took a pride in that they thought of themselves as valued skilled or semi-skilled labour. They came to work and they were very conscious if there was a fault on their particular part of the line, they were safety conscious. That has been destroyed – they can be deployed on any part of the line now, working alongside contractors all the time who are often poorly trained, don't have a safety culture and basically don't give a shit. That obviously demeans and undermines. When you put that together with macho management handing out disciplinary Form Ones like confetti it's all stick, very little carrot.[2] It's not a particularly pleasant or decent job, as the railway always was. It was never the best-paid job, but it was a safe job, in many respects it was a job that people took a pride in. It's gone.

Insecurity took on different forms elsewhere in the new railway. As mentioned above, the work of the non-driving operating grades in many of the train operating companies was contracted out or cut altogether. Managers within the industry have actively used the uncertainty created by privatisation to unsettle employees. Many workers felt insecure on a general level, but on an individual level what has made it worse has been the brutal way in which people have often been forced to leave the industry. The RMT spokesman I interviewed (1998) noted:

> One of the media-friendly people that came into the industry, Richard Branson – his Virgin companies, are one of the worst employers that we deal with. They are significantly worse now than the infrastructure companies ... they undertook basically a destaffing of the system initiative. Traditional jobs done away with, replaced by about a third of the people. It's very much corporate image, you know smart American-style uniform, having a very subservient role and not an important operational role on the railway. Get rid of as many staff as possible, decimate your personnel function. People go to work, expect decent conditions, a job, many instances they have got 24 hours to leave their desks – that's the world of Richard Branson on the railway.

Thus privatisation has achieved the aim of introducing pressure for change by means of fragmentation and the introduction of market forces. For most sections of the railway workforce, there is little in the way of long-term job security as labour costs are cut and conditions deteriorate. But this 'success' has not been won without creating new problems for the industry over the short, medium and long terms.

The contradictions of culture change

A tension exists between those who write on culture change. On the one hand there are the managerialist writers who argue for the possibility of, and in some cases the necessity for, culture change (see for example Peters and Waterman, 1982; Peters and Austin, 1986; Deal and Kennedy, 1988; Peters, 1992, 1995; Champy, 1995). In this genre, organisational culture is simply another variable that management can control and shape to ensure corporate success. At the same time there has been a growing critique by sociologists and anthropologists of some of the assumptions made about organisational culture in the management 'guru' literature (see Lynn-Meek, 1988; Willmott, 1993; Wright, 1994). As argued above, these academics point to the misunderstanding and manipulation of social scientific ideas on culture in such interpretations, the assumption of unitarist conceptualisations of the workplace, and the view that managers can unproblematically devise, change or manipulate culture for their own requirements. In the context of the railways, culture change initiatives, coupled with organisational restructurings have, to a limited extent, 'changed the culture', if we use the term culture in its broadest sense. If we think back to the notion of the frontier of control, we can see that the management tactics in such programmes are – whether consciously or not – to renegotiate the order within the employment relationship. In part this is achieved simply by removing some of the older and more experienced staff and by changing job content or procedures, thus affecting relations between workers. But this 'success' is not without contradiction or unintended consequences, and it is these that I now want to focus on in three areas:

1. skill
2. commitment
3. the greening of work.

Skill

Perhaps the most important area in which culture change programmes and organisational restructuring have occurred is that of skill and the corrosion of railway-specific knowledge. The process of privatisation, and the restructuring that led up to it, damaged the skills base in the industry in a number of ways. It is essential to see skills as not simply formal qualifications, but more broadly to include the tacit skill and knowledge that was embedded in the industry. Some have called this railroad or railway culture, and it was often described by my inter-viewees as 'railway sense'. As we have seen, the process of privatisation led to the loss of some 40,000 staff, and many of these were amongst the most experienced workers in the industry. But in addition to these losses, privatisation also produced a fragmentation that has had im-portant implications for skills. When the privatised railway was being designed in the early 1990s, there was a desire to create as much com-petition as possible in each of the many sections of the industry, to correct the perceived failures of some of the earlier denationalisations. This aim resulted in the creation of over 100 new units to which staff were allocated on a functional basis. The process was deliberately designed to create horizontal as well as vertical separations between the units – command and control replaced by competition and contract. The presumption on the part of the Government and their advisors was that these individual units would inherit the core com-petencies and expert knowledge of the BR workforce and there would therefore be no loss of skill. But this did not happen. As John Edmonds, the first chief executive of Railtrack, acknowledged:

> The third problem – the management of Railtrack's contractors – arose directly from the privatisation process, which had assumed the principle of the 'competent contractor'. In fact, many skilled staff had moved on during the various regroupings which had taken place during the sale process and Railtrack, following a series of in-cidents, had to step up the arrangements for the supervision and monitoring of its contractors and the registration of personnel com-petent to work safely on the network.
>
> (Edmonds, 2000: 67–8)

The need to formalise the system of training has been repeated across the industry and is also a feature on the London Underground. Essentially this formality is a consequence of viewing labour as an individualised set of competencies rather than seeing skills and knowledge as collective property. As we have seen in earlier chapters, formal training has not been the main route by which railway workers learnt their jobs. Rather 'railway sense' was transmitted through workplace culture and built up over a career. For management this had both positive and negative features. On the one hand the industry's training costs were relatively low as formal training was kept to a minimum, but on the other this gave workers autonomy and power. In essence this knowledge about work and the way it is transmitted became an integral part of railway culture. The fragmentation of the industry coupled with the removal of a large proportion of more experienced workers damaged this unacknowledged system. To some extent the problem of skill loss, both formal and tacit, has been disguised as more experienced workers 'look out' for the less experienced. Many experienced workers I interviewed openly worried about what would happen when this safety net was removed, either by further redundancies or early retirements.

Commitment and identification with work

A second area where culture change programmes in the railway industry have had an important impact is that of worker commitment and identification with either company or industry, long a feature of the rail industry in the UK and the rest of the world (see above and also Stein, 1978; Ferner, 1988; Gratton, 1990). Ironically all the culture change programmes pursued in the industry have explicitly attempted to promote commitment. The main problem for many workers, and by extension the industry as a whole, has been the chronic insecurity that has been deliberately injected into the system by privatisation or greater levels of commercialisation. Again part of the problem is the horizontal and vertical fragmentation of BR, which has seriously curtailed promotion prospects for workers. Historically, integrated railway companies have been large enough to offer alternative employment in different grades or in new locations for those seeking promotion or made surplus by job losses (Revill, 1991; Savage, 1998). Workers who moved in this way retained seniority rights as well as other benefits. Whilst not fully compensating employees for the disruption in their lives, such a system did much to smooth restructuring, particularly during the 1960s (see Bagwell, 1982). As the RMT official I interviewed in 1998 explained:

> There has always been an element of the industrial gypsy in the railway, particularly since Beeching in the early '60s when whole

lines were closed. A move used to be looked on as a long-term thing. 'We will move from Galashiels because there is no railway here anymore, we will go to Leeds or Edinburgh or whatever and we will move our house and do the full thing.' And they are still there 30 years later.

In the post-nationalisation era, such flexibility is severely curtailed in several ways. First, employees find themselves in far flatter organisations, which restricts movement. Second, there is less scope for most workers to move between different areas and grades, as this would entail leaving one company to join another, with a subsequent loss of employment rights (though the situation for drivers has been somewhat different, as discussed above). On top of the uncertainty associated with this fragmentation has been the insertion of temporal instability as a result of fixed-term contracts between the new elements of the industry. The sense of anxiety is compounded by the short-term nature of the franchises for the train operating companies that now run passenger services. These franchises range in length from five to 15 years, with most being let for around seven years. At the end of the period a new company could win the franchise and decide to restructure and force staff to move again. The RMT official I interviewed (1998) acknowledged that:

> There is not a great deal of optimism in terms of their future railway careers. The franchises are for seven years but there could be six owners in those seven years. It's very difficult to plan your career if you are a railway worker these days; there is just no overall security.

This sense of anxiety was not simply an unintended consequence of liberalisation, but was rather intended as an inherent feature of the newly privatised system, a structure actively designed to destabilise the 'traditional' railway workforce (see for example Glaister and Travers, 1993; Freeman and Shaw, 2000).

On the London Underground there has also been a concerted effort to emphasise the insecure nature of railway employment. Jon Filby of the LUL employee development team wrote in an in-house staff magazine:

> the reality of the 1990s is that no company is able to guarantee a job for life to anyone ... People must understand that job security can still exist, but that in the new world that security will come

from within the individual not from the company. The secure employee will always ensure that they have relevant skills that are in tune with the requirements of the marketplace.

(*U Magazine*, 8 March 1995: 13)

This shift is important, rhetorically marking those who ask for *any* level of security as demanding 'a job for life'. Responsibility for career now becomes individualised, and anxiety is therefore a manifestation of individual failing rather than collective concern. The fragmentation and flux of the contemporary world of work become features to be actively embraced rather than hidden from.

As we saw above, the drivers as a group have done reasonably well from the fragmentation of the rail industry and the introduction of market forces. But this improvement in their basic conditions has done little in terms of commitment to organisations within the industry, and drivers are now willing and able to move between companies in a way other railway workers cannot. This ability to move to follow wages could be seen as a product of insecurity, in that drivers are no longer prepared to offer commitment to firms that have perhaps only a seven-year franchise, at the end of which a new company with new labour policies may well take over. As a spokesman from ASLEF (interviewed in 1998) told me:

One of the political ideas behind privatisation was to break up the idea of drivers as being a British Rail driver, didn't matter if you were in the north of Scotland or the south of England, you were a British Rail driver, and they sought to smash that and they have done it, but they have encouraged people to be more mercenary. If you are in the north-east of Scotland on £20,000 and if you don't mind where you live, you will move down to England for £25,000 – you've got no loyalty to your company, and why should he have? They have no loyalty to you.

The same official went on to discuss some of the implications for organisational loyalty:

One of the things that kept it [*the railways*] going was the loyalty factor I am sure. A loyalty to a nationalised industry where they thought they were giving a service to the public. They have been discouraged from having that loyalty. Companies now want the loyalty to Stagecoach or Seacontainers [two groups that have taken over train companies]. There is no loyalty to those companies.

A constant refrain from older workers I interviewed was the lack of commitment on the part of newer entrants, and many cited the lack of interest in the job and the industry. One signal worker in his forties compared younger workers with his own experience:

> I was proud of being a signalman, of my job, of being a Met man. I joined the railways because I was interested in railways, now people join the job because it's a job and it's quite well paid compared to outside industry. I talked to one young bloke and he said 'oh I just push the buttons and pull the levers'.
>
> (Huw Evans, LUL signal operator, interviewed 1995)

Another signal worker, who had joined in the 1960s, reflected on the experience of younger workers:

> It wasn't just a job to me, now it's just a job. Now people are just looking for a job and they'll do anything because there's just not enough jobs outside.
>
> (Fred Ross, senior signal operator LUL, interviewed 1995)

It would be wrong to dismiss these quotes as the nostalgia of older workers. What surprised me most during the research for this book was the way much younger workers reflected on work. There was a sense of this generation being the last of the 'true' railway workers. This was similar to the feeling that was apparent in the words of railway workers quoted from the 1960s (see Chapter 4). However, those workers were in their late fifties or early sixties in the 1960s, whereas many of the workers I talked to in 1994–6 were in their mid- or late twenties or early thirties. This is best illustrated by a LUL signal worker in his mid-twenties:

> It's getting worse. The main reason is there's a bunch of kids in the grade because of the mass exodus of the older geezers during the voluntary severance. A lot of people now have a very laid-back sort of attitude to the job. If it screws up, they couldn't care less. It's difficult to know how you can be proud considering the way it's going, years ago you would come in and look forward to doing the job. Now you look at your watch and want to go, especially now with the people you've got in the grade who don't give a toss about their grade and don't take any pride in it.
>
> (Simon Gilbert, LUL signal operator, interviewed 1995)

While these more established workers were critical of newer workers, their view is more complex than would first appear. Older railway staff *were* critical of a lack of vocational commitment by younger workers, while recognising that such workers had no reason *to* display such loyalty. Many older workers expressed a wish to leave railway service altogether because of this changed climate, despite a very real interest in, and commitment to, the industry. There are interesting parallels here with the findings of other sociologists who have looked at organisational restructuring, and in particular the way older workers more readily identify with companies. Both Casey (1995) and du Gay (1996) note the way older and more established workers retain a commitment to organisations in spite of being marginalised by culture change programmes. Casey in particular suggests that it is the older workers who have real loyalty and engagement with their companies and work. While there is some erosion of this commitment for older workers, this is in marked contrast to younger workers – those under 35 in Casey's example – whose 'bond to the company is tenuous and fragile' (Casey, 1995: 116). Among the younger group of workers in her sample Casey reports a far more instrumental attitude towards work: pay, locality, health insurance etc. are what count. Rarely did this group talk about other sources of allegiance to work, as older workers had done. Interestingly Casey argues that the difference in loyalty to the company displayed by these two groups cannot simply be put down to differences in length of job tenure – some of the members of the younger group had worked for the company for over ten years. Rather she suggests that what is being witnessed in the workplace is something far more profound:

> Attitudes are a matter of generational location and a change in the cultural production of loyalty and solidarity, a result of the erosion of industrial society and of the traditional, organic solidarities that provide cohesion.

> (Casey, 1995: 116)

In the context of the railway industry I think that the situation is more complex, as many of the workers I interviewed were in this same younger age group, and while there was undoubtedly a generational difference, many of the under-35s nonetheless identified with the older generation, considering themselves as 'proper' railway workers in contrast to the still younger and newer workers. Several workers in their twenties described the way they were seen as 'dinosaurs' as a result of their socialisation and expectations about work.[3]

The greening of brownfield sites?

Perhaps the ultimate contradiction for management engaged in culture change is the greening process itself. At the beginning of this chapter I raised the issue of the desire of management to change the architecture of the employment relationship by removing the potential oppositional resources that derived from occupational identity rooted in experience. What management object to is precisely the experience of the workers, and in particular the ability that they derive from that experience, the fact that they are so embedded in the labour process that they have power in determining how work is to be carried out. Such power can be understood both as an objective skill or knowledge over work, and also as a moral authority in the workplace that shapes the relationships between workers and between workers and their managers and supervisors. In the past this restriction on the power of management was further strengthened by workplace rules and agreements, which inserted formal bureaucratic structures that both empowered and constrained workers and management. As the American labour historian Jack Metzgar (2000: 36) reminds us in his discussion of the US steel industry:

> Today ... both the Left and the Right attack this 'workplace rule of law' with little sense of its role in liberating workers from arbitrary authority and all the indignities, the humiliation, and the fear that come with being directly subject to the unlimited authority of another human being.

To a limited extent this sense of control allows workers a degree of dignity and autonomy over their work and is a further mark of an adult identity. The achievement of adulthood on the part of the workers is won at the expense of lost managerial discretion and control over the labour process. The desire to change culture in this context is the wish to reverse this relationship, to reduce the workforce collectively to the status of adolescence or even childhood. The problem, however, for management is precisely this: social relations or organisational culture are not static but a *process* taking place within a context. Even the youngest child grows up eventually.

If we place this situation within the framework developed by Mead (1978), we can see a series of contradictions for management. The process of 'greening' is understood as the attempt to create a cofigurative culture, wherein the young become the role models for the younger. This is in contrast to traditional societies, in which the older generation provides such guidance. Mead views such configuration as a feature of those societies that have undergone shock, or have been displaced, where the knowledge of the elders is of little or no relevance to the younger generation or may simply be entirely absent. Importantly, Mead goes on to suggest that configuration will normally last only one generation, as cultures revert back to post-figurative form after the singular circumstances that created change have passed or become normalised.

Thus post-figurative cultures must be seen as re-emergent processes. The 'one-off' gains that can be made by management as part of a culture change programme are fragile in the face of changing market situations within a dynamic capitalist economy. Thus the 'one-off' change that solves all the organisation's problems is probably but one in a series (see Ramsay, 1996; Strangleman and Roberts, 1999). As Sennett (1996: 8) points out, to believe that you can plan and shape so precisely something as complex as 'culture' is fundamentally flawed:

> For in this projected future there lies a way of denying the dissonance and unexpected conflicts of a society's history. This attitude is a means of denying the idea of history, i.e., that a society will come to be different than it expected to be in the past.

The potential re-emergence of opposition and resistance within a 'green' workplace is an issue that management are obviously aware of, and one option seems to be the celebration of permanent frenetic change, conveniently matching the corporate form to that of its markets in the 'nanosecond nineties' (Peters, 1992) or 'the rough weather we're sailing in today' (Champy, 1995: 76). O'Connell Davidson (1993: 155), for example, quotes a manager from the water industry extolling the virtues of mixing the company's supply of labour by using in-house staff and external contractors:

> I don't think to a limited degree a bit of uncertainty is a bad thing, that they're not entirely sure that they are secure, so that they work harder to make sure the company goes.

A similar sentiment is quoted by Storey (1992: 151), who reports a conversation with Geoff Armstrong, research director of Metal Box and chair of the CBI employment committee:

> He propounded the view that because of the tendency for old habits to re-establish themselves and for complacency to creep back in

rather rapidly, there was merit in engendering an almost permanent sense of crisis. This could be engineered through the timely release of information, through periodic reorganization and through a constant flow of initiatives.

But the problem for management determined on such a course is the loss of knowledge and, crucially, flexibility, within the green workforce. As Mead (1970: 39) points out:

> The new culture often lacks depth and variety and, to the extent that it does ... may be less flexible and less open to adaptive change than the old postfigurative culture was.

The issue of adaptability, or the lack of it, emerges in Marris's (1974) exploration of the conservative impulse in individuals and societies. In his defence of conservatism, Marris is not arguing for a static notion of social order; rather it is the *interaction* of two contrasting qualities that is important for adaptability and flexibility. The ability to cope with change, therefore, is dependent on the friction *between* confidence, in the sense of the regularity and predictability of social behaviour, *and* the ability and willingness to revise both purpose and understanding. As Marris emphasises, the predictability of events is rooted in knowledge and experience, and such experience is a mark both of the adult individual and of the social group. In *Uses of Disorder*, Sennett (1996) suggests that the search for purity is essentially an exercise in filtering and excluding social experience. Unity, whether it be in a suburb, teenage gang or workplace, does violence to a rounded, complex and more adaptable human experience.

The attempt to 'green' a workforce is, therefore, riven with contradictions. Firstly, the green workforce is one that may be more pliable, but this is achieved at the expense of flexibility. Secondly, the removal of a critical mass of experienced workers may result in corporate memory loss, with the consequence that both formal skill and informal, tacit knowledge about the organisation may be destroyed or seriously damaged. Thirdly, if we understand social relations to be a social process, then the greening process is likely to be reversed over time as the younger and less experienced workers gain their own knowledge.

Conclusion

This chapter has been about the contestation of the past. In many ways the quotes from railway workers interviewed for this book appear very similar to those used by sociologists in the 1960s and early 1970s that were examined in Chapter 4 (Salaman, 1974; Hollowell, 1975). We have here railway workers seemingly lamenting the passing of a golden age of work, one that has been destroyed by management and as a result of the recruitment of new staff who are not 'proper railway workers'. But is this simply another instance of nostalgia prompted by the turmoil of restructuring and privatisation? As Davis (1979) and others have argued, nostalgia tells us more about present conditions than it does about the past. The railway workers quoted here have undergone a far more traumatic process than their counterparts in the 1960s, with their industry experiencing massive job losses coupled with the fragmentation of the railway vertically, horizontally and temporally. Insecurity is now felt in restricted promotional prospects, a limiting of sideways movement within grades and the short-term contracts that have become a major feature of the industry. If reflective nostalgia is a feature that unites workers across the decades in the industry, then what is different about the voices from the 1990s is the way they include many younger workers whose views are similar to those of their older colleagues.

If there is simple nostalgia here it resides not in the workforce, but rather in the desire of management and politicians to return to an idealised version of free markets. In many ways the impetus behind culture change, commercialisation and, ultimately, privatisation, has been nostalgia for an ideal railway worker of the past – one loyal, dependable, respectable as well as autonomous. At other times this quest has taken on a more modernising aspect with a utopian desire to create a new kind of workforce, one that is engaged, dynamic, entrepreneurial and committed. These visions of the ideal worker have much in common, as one of the features that unites both utopians and nostalgics is the fact that their pure visions are shorn of the awkwardness of everyday reality – these are images untainted by complexity. As John Carey (1999: xii) wrote in his essay on utopian thought:

> The aim of all utopias, to a greater or lesser extent, is to eliminate real people. Even if it is not a conscious aim, it is an inevitable result of their good intentions. In a utopia real people cannot exist, for the very obvious reason that real people are what constitute the world

that we know, and it is that world that every utopia is designed to replace.

While this quest for the ideal may be flawed, many of the new companies formed by privatisation have nonetheless attempted to change the culture, and to some extent have enjoyed success. In the final chapter we examine some of the dramatic and unintended consequences of this process.

7
Nostalgia for Nationalisation?

> Our efforts to convert this business to a customer-first culture
> is a work in progress.
> (Philip Mengel, chief executive of English, Welsh and Scottish
> Railways, *Rail*, 21 February 2001)

> The collapse in professional delivery has been the biggest sur-
> prise of rail privatisation. Simple things that railway people
> once did without thinking have now become a major crisis.
> (Chris Green, head of Virgin Trains, *Rail*, 21 February 2001)

At a little before 12.30 pm on 17 October 2000, a Great North Eastern
Railway express travelling from London to Leeds crashed just outside
Hatfield in Hertfordshire. The cause of the accident was a faulty rail,
which shattered into more than 300 pieces as the train passed over it.
Four people lost their lives and many more were injured. As the full
implications of the crash became apparent, Britain's railway system
nearly ground to a halt. Railtrack imposed speed restrictions on 1,200
sections of the network after a further 3,200 cases of Gauge Corner
Cracking (GCC) – the immediate cause of the crash at Hatfield – were
discovered. Some journey times were quadrupled, Scotland was cut off
from England for a time and the media was dominated by sensational
stories of disgruntled passengers and railway company incompetence.

In many ways the fragmenting of a length of rail in the English
Home Counties acts as a powerful metaphor for privatisation itself: a
seemingly solid structure held together out of habit but in reality a collection of hundreds of fragments, the faults of which many had
warned of long before. The incident, and the way it was subsequently
handled, tells us a great deal both about privatisation and the com-

plicated structures created by it. The track and infrastructure at Hatfield were owned by Railtrack, which subcontracted maintenance to other companies, most directly the civil engineers Balfour Beatty and Jarvis Fast Line. In turn, some of this work was subcontracted to one of the 2,000 companies now operating in this sector. These firms had to employ the rail freight company English, Welsh and Scottish Railways to deliver the replacement rail to the site, something that had proved difficult to accomplish over several months before the crash. Trains from three rival passenger train companies – GNER, West Anglia Great Northern and Hull Trains – passed over the track at Hatfield every day. Each of these companies hired their rolling stock from one of the three rolling stock leasing companies (ROSCOs), which in turn looked to others to maintain them. Train operators in competition with one another each paid Railtrack for access to the network in a complex system of contracts and agreements.

Though the crash of the GNER train at Hatfield was caused by a defective rail, it was clear to many that the privatisation of British Rail was the underlying reason. At the heart of the problem lay the fragmentation of the industry and the complex interrelations between parts of the new structure that produced competition and conflict rather than co-operation. The state of the track at Hatfield had concerned engineers some time before the incident, but attempts to carry out remedial work were hampered by problems in obtaining possession of the line when trains were not running. Railtrack was reluctant to allow workers from Balfour Beatty or Jarvis to replace track as this would have meant delaying or suspending train services, which in turn would have caused the train operating companies (TOCs) involved to have incurred penalty payments. The TOCs were concerned to minimise any inconvenience to their passengers, as well as being wary of the possibility of having fines imposed on them by the Rail Regulator for poor service. Therefore the simple issue of damaged rails, a routine aspect of railway operation, became far more serious than it need have been. Under BR there had always been a straightforward hierarchy and clear lines of responsibility for all aspects of railway operation. With privatisation and the insertion of market logic at every level and part of the industry, this structure was replaced by a nexus of contracts, key performance indicators and a highly complex system of fines. Where lines of communication had once been unambiguous and intelligible, they were now blurred and decision-making overly bureaucratic.

What became known simply as 'Hatfield' brought the railway industry and the nature of its privatisation under close scrutiny, and faith in market-based solutions for the problems of the public sector was challenged. The chaos that ensued in the autumn and winter of 2000 was compared unfavourably to World War II conditions and in particular the spirit of the blitz, when 'the railways always got through'. Enthusiastic journalists, warming to their subject, took great delight in reminding readers of the wartime posters in which potential passengers were asked 'Is your journey really necessary?'. When the crisis was deepened by extensive flooding following the torrential rain of that autumn, the railways became symbolic of a crisis in national identity: the country that invented the railways was unable to offer even a third-world service.

Nostalgia for a lost workforce

Interestingly much of the criticism of the industry after Hatfield focused on the lack of a skilled workforce capable of identifying the problem before the crash or coping with its aftermath. The *Financial Times* (22 February 2001) highlighted the issue of railway labour, suggesting that job insecurity (discussed in Chapter 6) led to a situation in which:

> The first consequence was the breakdown of the old comradeship, which used to mean that the problems were easily spotted, repairs made, and people could talk to each other. Track workers operated in gangs and knew their stretch of rails like their own back gardens. Instead, workers became nomadic, moving to the next job with little or no local knowledge and instructions not to talk to rival workers.

In December 2000 Railtrack had moved to restore a measure of public confidence, as well as skill in the industry, by announcing the recruitment of 1,000 new workers to fill signalling and engineering posts. Amongst these the company explicitly targeted recently retired railway engineers made redundant before, during and after privatisation. On 5 May 2001 the *Guardian* reported under the headline 'Railtrack asks army for help' that the company was so desperate for skilled staff that they were attempting to hire engineers from the armed forces. The article went on to record that Italian, German, Indian and Romanian rail engineers were also being sought, and noted:

> Railtrack's problems began after privatisation when it lost many experienced members of staff who realised that the company was bent on sub-contracting out most of its engineering work.

Much was made of the incompetence of senior figures in Railtrack. At the time of Hatfield just two of the 14-strong Railtrack board had a railway background. This privileging of finance capital at the expense of engineering knowhow was perceived as further evidence of industrial decline. The chief executive, Gerald Corbett, had previously been a finance director of the drinks group Grand Metropolitan, and had brought in new management which the *Financial Times* (21 May 2001) suggested 'became symbolic for many railwaymen of the new culture of bureaucrats and accountants'. The same article went on to report that Corbett had boasted at a lunch of having changed 60 of the top 100 people during his time at Railtrack. Within the industry, Hatfield, and previous crashes at Ladbroke Grove (1999) and Southall (1997), finally allowed more experienced managers to voice doubts about the faults of privatisation. Chris Green, the head of Virgin Trains and a career railwayman who began in the industry as a BR management trainee in the early 1960s, wrote:

The result has been a collective loss of memory on the basics of running a railway. Old skills have not been valued and experienced staff have been dismissed in ill-considered cost-cutting initiatives.

(*Rail*, 21 February 2001)

These and other accounts of the loss of skill during privatisation were pre-figured by *Guardian* journalist Jonathan Glancey. In a sentimental piece in 1999 reflecting on the opening of a new annex to the London Transport Museum, he lamented the demise of the railway work ethic, suggesting that it, too, was now only fit for the museum. Comparing contemporary workers with past generations of train drivers, he suggested:

Such men were like champion jockeys in charge of mechanical race-horses looked up to by even the most expensively educated British schoolboy ... More importantly, they were well read, dedicated union men, who learned their craft, as well as the ideas of Ruskin, Morris and Marx, in evening classes and in oily depots. They were truly the aristocracy of labour. Not today. Those working for the privatised transport companies run by former supermarket managers and second-hand car salesmen have become little more than the 'lump'.

(*Guardian*, 22 September 1999)

Nostalgia for shareholder value

Both Railtrack and the Major Government had made much of the private sector's ability to raise capital to fund the renewal and expansion of the rail network without adding to the Public Sector Borrowing

THE MOMENTOUS QUESTION.

"Tell me, oh tell me, dearest Albert, have you any Railway Shares?"

Figure 7.1   The Momentous Question. Queen Victoria asks her husband Albert if he has any railway shares in the wake of the bursting of a speculative investment bubble in 1845. (*Punch*, 1845, Vol. 9, p. 45)

Requirement (PSBR). After Hatfield, however, the ability of Railtrack and the rest of the rail industry to fund investment was called into question. At the time it was privatised, Railtrack's finances were engineered in such a way as to give it an 'AA' credit rating (see Edmonds, 2000: 61). This meant that the infrastructure company was an attractive investment and, more importantly, it could borrow money from the markets more cheaply because of the relative lack of risk involved. At flotation in 1996, the company's stock had been valued at £3.90 a share, but this steadily climbed to a high of £17 during 1998. After the three crashes at Southall, Ladbroke Grove and Hatfield and the gradual exposure of the potential liabilities that the company faced, its share price plummeted to under £4. This collapse in confidence on the part of traders and individual shareholders weakened Railtrack's ability to raise finance at a time when it was being exposed to new liabilities on a daily basis. In June 2001 Railtrack tumbled out of the FTSE 100 index, and financial analysts at city firm ABN Amro warned that even at this level Railtrack was highly overvalued. Given its exposure, a more realistic share price according to ABN Amro was 58p (*Financial Times*, 7 June 2001). In response to these events the company made it clear that it could no longer be expected to invest at the level it had intended, and it began to ask for assistance from the Treasury simply to finance the repairs and renewals it was currently undertaking.

Nostalgia for nationalisation

Railtrack's perilous finances neatly mirrored the public perception of the industry more widely, and stimulated a growing chorus in the press for the return of State control. Some of these calls were from expected quarters such as the centre-left *Guardian*. Barely a week after Hatfield the paper had published a piece entitled 'Return ticket to state control easily bought' (*Guardian*, 23 October 2000). Later Jonathan Freedland penned 'Ready to renationalise' (*Guardian*, 14 February 2001), and a leader on 26 May 2001 was headed 'Re-engineering Railtrack: it must be returned to the public sector'. Similar opinions were expressed across the political spectrum. The *Financial Times* (7 June 2001) reported that some of Railtrack's largest shareholders wanted to see a partial renationalisation of the company as a way of boosting con-fidence in the sector as a whole. Similarly John Kay wrote a piece sym-pathetic to State control headed 'Get back on the rails. Privatisation has failed. It is time to consider nationalising Railtrack' (*Financial Times*, 27 June 2001). But perhaps most surprising of all was that the right-leaning *Daily Express* labelled privatisation 'a disaster ... leaving the nation with one of the worst rail systems in Western Europe' (*Daily Express*, 14 March 2001).

The end came over the weekend of 6–7 October 2001, when Transport Secretary Stephen Byers refused to extend Government funding for Railtrack and so forced the six-year-old company into receivership. Without further advances from the Treasury, Railtrack was bankrupt. Its share price was effectively reduced to zero, making the earlier valuation by ABN Amro of 58p per share seem wildly optimistic. While the Government studiously denied that this was 'renationalisation by the back door', there can be little doubt that the State had regained control of part of the industry very cheaply. Had the Government chosen to purchase the shares before the events of October it would have been forced, under European law, to compensate shareholders on the average price of the stock over the previous three years. This would have valued each share at nearer £8 or £9 rather than the trading price of £3.60 at the point of collapse. Predictably the Railtrack board fought hard for 'proper' compensation for their shareholders, claiming that its troubles were due to Government duplicity rather than corporate mismanagement.

Paying the price for privatisation

This seems an important point at which to take stock of the privatisation of the rail industry, before going on to examine its wider implications. There can be little doubt that the privatisation of BR has been a costly failure. The industry still requires massive public subsidy in order simply to survive. But the difference now is that the flow of funding around the system is far more wasteful and bureaucratic than it ever was under BR. This price has been paid in the restructuring of the industry in the lead-up to privatisation, the redundancy pay, the fees of an army of lawyers, accountants and other consultants who have profited from the industry's break up. Far from freeing the taxpayer of a financial burden, privatisation has led to the writing-off of hundreds of millions of pounds of debt in order to make individual units attractive to investors.

If privatisation was financially flawed, it has also exacted a heavy human cost, most obviously in the crashes at Southall, Ladbroke Grove and Hatfield. Many argue that passengers and railway workers alike were killed as a direct consequence of the structure put in place at the time of privatisation. But there has also been a human cost in a less direct sense, borne by the thousands of current and former workers who have suffered in the lead-up to, during and after the fragmentation of, the industry. Many workers have simply left the industry as a result of the early retirement packages on offer, while others have been made redundant. For those who have stayed, job security is limited and conditions have deteriorated. While deaths in accidents were unintended consequences of the privatisation process, the working conditions of current employees were an intentional result. Privatisation explicitly aimed to introduce competition into the railways, and in order to achieve this the unitary BR was divided horizontally and vertically, replacing the command and control structures with a myriad pattern of contracts. The belief was that the customer/provider split would act as a disciplining device on both employees and management, developing entrepreneurial skills in both groups. In addition to this horizontal and vertical functional fragmentation, there has also been a deliberate injection of temporal insecurity with the use of fixed-term contracts. The combination of these types of fragmentation has produced new levels of insecurity and has also corroded much of the skill and tacit knowledge of the workforce – both blue and white collar. Complex patterns of social relationships, which were a feature of the industry for generations, have been disrupted or entirely removed, and basic principles of railway operation have been lost, in some cases for good. In his report on the Southall crash, John Uff (2000: 208) focused deliberately on the question of culture within the privatised industry:

> The railway industry is now overburdened with paperwork, such that it is to be doubted that many individuals can have a proper grasp of all the documents for which they bear nominal responsibility. The stock answer to any problem which is identified is to create yet more paperwork in the form of risk assessments, further Group Standards and the like ... No primary or secondary paper-based system is a substitute for common sense and commitment to the job.

When the politicians in the Major administration or the new owners of the industry were warned by experienced railway workers of the potential problems, their advice was casually dismissed as opposition from time-served bureaucrats unused to the discipline of the market. As John MacGregor, one of the key architects of rail privatisation, announced to a parliamentary transport select committee, publicly owned transport systems:

> could not hold a candle to the market and competition as the best ways of determining what the travelling public want.
>
> (MacGregor quoted in Wolmar, 2000: 132)

Both politicians and some of the new players in the industry ignored the inconvenient aspects of railway history, choosing instead to rely on a caricature of the past. In this reading the railways emerged in the heroic era of British capital, constructed by private shareholders and risk-taking entrepreneurs, unfettered by the dead hand of the State or petty regulation. Such a picture both lionises the role of the entrepreneur, and also allows space for a more benign view of the railways as a key moment in the construction of a timeless rural idyll bringing order and intelligibility to the countryside. In turn this tells us much about the tensions in Conservative thought between libertarian free-marketeers on the one hand, and social conservatives on the other. The strains in this relationship were always apparent throughout the Thatcher years, and continued to bubble under the surface when John Major became Prime Minister in 1990. Rail privatisation seemingly resolved these tensions by striking at one of the last nationalised industries, towards which both wings of the Party were hostile, while seeking to restore or rediscover a lost era – that of a nation at one with itself.

Privatisation at the end of the line?

So what does the story of commercialisation and privatisation in the rail industry tell us in a wider sense? The sale of BR was the last gasp of the privatisation process in the UK, John Major's big idea for his second full term. The process and the structure chosen for the railways were in many ways a response to the criticism and obvious faults associated with earlier sales of State assets. Most marked of these criticisms had been that far from injecting market discipline into the moribund nationalised industries, the privatisations of the 1980s had merely substituted a public-sector monopoly with a private-sector one. In the case of rail, ministers and their advisors, particularly those from free-market think-tanks, argued that market forces should be allowed to play a fuller role in shaping the industry. For all the Conservative rhetoric about being the pragmatic party of business, the process of privatisation in general and their unalloyed faith in the market in particular are touchingly romantic, utopian and even nostalgic.

Writing in her reflections on the 1960s, Sheila Rowbotham (2000: 1) talks of the disjuncture between memory and interpretation in the evaluation of the decade. As she puts it:

> It has arisen because, as the hopeful radical promise of the sixties became stranded, it was variously dismissed as ridiculous, sinister, impossibly utopian, earnest or immature. The punks despised the sixties as soppy, the Thatcherite right maintained they were rotten, the nineties consensus was to dismiss them as ingenuous. Dreams have gone out of fashion, making a decade when they were very real appear incongruous and elusive.

It seems to me that this analysis of the way in which the 1960s is remembered and subsequently used in political discourse actually applies to a much longer period. The whole of the era of the postwar consensus could equally be seen as a space that has variously been seen as 'soppy' or 'rotten'. Nationalisation, and more widely the public sector, have become caricatured as having failed – a nanny State that tried to reach into every aspect of life, to interfere and control private and commercial activity. Raphael Samuel, in his essay 'Mrs Thatcher and Victorian values' (1998: 344), writes:

> The Welfare State, under this optic, appeared as Old Corruption writ large, a gigantic system of state patronage which kept its clients in a state of abject dependence, while guaranteeing a sheltered existence for its officials and employees.

This periodisation allows the time before nationalisation to be liberated from any awkward negative associations. The Victorian economy becomes a model to follow, and *laissez-faire* a political Holy Grail. Economic success can be realised through the application of hard work, entrepreneurial flair, minimal State intervention and the operation of the free market. Politics, therefore, becomes an exercise in restorative nostalgia, mixed with a healthy disregard, or nostophobia, for the more recent past.

This faith in, and vision of, the market have of course been adopted across the political spectrum. New Labour under Tony Blair have enthusiastically embraced the market with the zeal of many converts to a new cause. Here again there is an intoxication with both the market and business, ironically just at the point where the limits of market logic and its effect on public services become daily more apparent. The private sector is associated with dynamism, with change and efficiency, whereas the public sector is associated with failure and poor service. Like the Conservatives before them, New Labour seem happy to promote this polarised, deeply ahistorical account in all aspects of their programme from schools to hospitals, air traffic control to, most recently, London Underground.

New Labour have, of course, a vested interest in repeating and building on the mantra of the failure of the postwar consensus in general and nationalisation in particular. Blair's interpretation of the past failure of the Labour Party and the wider movement is inextricably linked to the idea that State intervention and control of the economy were misguided. Like the Conservatives before them, the 'winter of discontent' of 1978–9 is a useful shroud that the Labour modernisers can wave at those who hanker after greater State involvement in industry, or who question the increased role of the private sector in the public realm. Grainy newsreels of the late 1970s add historical distance between the modern 'now' of the market and a less black and white world of the mixed economy.

But there is another interpretation of this history. This book has, in part, been tracing different interpretations and constructions of the past. While it has focused on narrating the story of the move to a greater commercialisation of the railways and eventually their privatisation in the 1990s, it has also had the aim of re-evaluating the era of nationalisation that started in 1948. The programme of State control after World War II was a revolution in economic and political policy and yet at the time seemed, if not inevitable, then a necessary course of action. Clement Attlee's nationalisations owed far more to pragmatism than to ideology, in sharp contrast to their eventual reversals. In 1945 British industry was in need of massive restructuring and investment. Only the State could provide the capital and authority to do this. In the case of the railways, the war had taken a tremendous toll on the industry's infrastructure, but its problems – which it shared with other sectors – ran far deeper. The railway industry had been established in an era when economic liberalism was the dominant political discourse. It developed in a fragmentary and piecemeal fashion, structured by competition and duplication, which prevented efficiencies being made through economies of scale. In the wake of the Great War there were moves to consolidate the system, with the Grouping of 1923, and while this had some success, the industry's problems remained mani-

fest. It was only with nationalisation in 1948 that many of the problems of the industry were finally addressed: overcapacity was ruthlessly
stripped out, lines were closed and investment was made in new technology in a way that would have been impossible economically,
socially or politically under private ownership. These efficiency gains
were won against the background of political interference, chronic
short-termism – especially with regard to finance – and with little in
the way of industrial unrest.

The study of the nationalisation and subsequent privatisation of the
rail industry also tells us something about the contemporary nature of
work and the way it is managed. While the cultural turn in management thought may be a relatively recent phenomenon, it is perhaps
more useful to see it as being part of a much older tradition in which
corporate failure is blamed not on structural factors but on the 'weakness' of the workforce. As Theo Nichols (1986) noted in *The British
Worker Question*, such an interpretation can be traced throughout the
nineteenth century and into the twentieth. Simplistic 'cultural' explanations that focus on employees, particularly those in the public
sector, are popular among politicians and managers. The need for a
'new culture' both sanctions change in the way work is organised, and
also offers a ready-made antidote to those in an established workforce
who oppose management. Awkward questions can be ignored as the
predictable last gasps of nostalgics, of industrial dinosaurs unable to
live in the modern world who attempt to hang onto the security of the
past.

Allied to, and intrinsically linked with, culture change initiatives
have been far wider shifts in the nature of work in Britain and the rest
of the world, such as new work practices, the implementation of new
technology and the general speed-up of organisational change and corporate fragmentation. In *The Corrosion of Character*, Richard Sennett
(1998) discusses the changing nature of work in what he labels 'the
new capitalism'. He suggests that while capitalism has always been
brutal, for many it was possible in the past to find meaning in work
because the system was more intelligible, more legible. In the contemporary world the speed of change and the ways in which employment
is being fragmented at once render work less legible and make its
effects more personal in their consequences.

This situation is celebrated by neo-liberals, not least because of
its disciplinary impact on labour and management. But at the same
time there are important contradictions in such a regime. This type of
environment, of which culture change programmes are simply one

manifestation, is corrosive of the workforce's commitment to their work. By this I mean not simply the day-to-day willingness to work, but rather a more profound sense in which people engage with their jobs and find meaning in them. What management often object to in established workforces is that they display a moral ownership over their work and, by extension, over the way it is organised. This embeddedness is the result of sedimented individual and collective experience over time. When change is arbitrarily imposed or enacted on such a workplace, it is precisely *because* of the commitment of the workforce, *because* they have invested meaning in their work, that there is resistance. Conservatism on the part of the workforce towards change is therefore an essential part of what it is to be adult – namely that people use their experience of the past to guide their present and future actions and attitude. In actively targeting the established part of a workforce, in trying to engineer adolescence, management are effectively destroying the quality of commitment to work. In the process they damage wider social relations, and the tacit skill and knowledge that organisations need.

In the conclusion to his report on the Southall crash, John Uff (2000: 202) makes the following observation:

> The lesson to be learned seems to be that compliance with Rules cannot be assumed in the absence of some positive system of monitoring which is likely to detect failures. Such a conclusion would, however, be a sad reflection on a fine industry which has been created through the enthusiasm and support of countless individuals who were proud to be thought of as part of 'the railway'. Perhaps the true lesson is that a different culture needs to be developed, or recreated, through which individuals will perform to the best of their ability and not resort to delivering the minimum service that can be got away with.

Perhaps there is a need for the kind of courses that Clive Groome teaches in his company Footplate Days and Ways (see Chapter 1) after all.

Epilogue

On 3 October 2002, the operational responsibilities of Railtrack plc, i.e. its track, signalling and station responsibilities, were taken over by the new, not-for-profit, Government-backed company Network Rail. Railtrack Group was left as little more than a shell company, with £1.3 billion in cash. Taxpayers are paying approximately £2.37 per share in compensation to Railtrack shareholders. Railtrack Group plans to seek voluntary liquidation.

In an interview with the *Financial Times* (3 October 2002), Iain Coucher, deputy chief executive of the new Network Rail, said:

> We are putting a challenge down to the people of Railtrack. What's happened in the past is the past.

Notes

Chapter 2

1 All tons are imperial tons, unless otherwise stated.
2 Parliamentary Select Committee on Railway Servants (Hours of Labour) of 1891, volume xvi: 1–91.
3 The term 'passed' refers to the fact that the individual cleaner, fireman or driver was passed, or qualified, to undertake the work, but was not doing that work all the time. Only after a certain number of turns or shifts (over three hundred) actually doing the work would the worker be fully rated at the new status. At certain times, during wartime especially, the new rate would be won quickly, but in times of slow promotion it could take many years.
4 All interviews were carried out by the author.
5 The 'road' here refers to a particular route. A qualified driver would have to 'learn the road' before he could drive a train over that section. In particularly complex locations this could take a considerable time. Semmens (1966: 69), in his biography of Kings Cross engineman Bill Hoole, describes how learning the five miles from the London terminal took between two and three months and involved detailed knowledge of 1,000 individual signals as well as hundreds of other features and peculiarities.
6 Before proceed signals can be given, the signal worker must 'set the road'. Points and signals are interlocked to prevent points being moved while a signal is displaying a clear aspect. In semaphore signalling, signals would sometimes fail to clear even though the correct lever had been pulled.

Chapter 4

1 The terms 'power box' and 'power signalling' refer to methods of moving or changing points and signals using power switches or levers, as distinct from traditional mechanical signalling, in which the signal worker manipulates points and signals by levers, using his own strength. In the case of power boxes, control rooms can be remote from the site controlled and it is possible to concentrate vast areas into one location.

Chapter 5

1 I use LU (London Underground), LUL (London Underground Limited) and LT (London Transport) interchangeably in this book, mainly because the staff interviewed here do.
2 Loadhaul was the company formed out of Train Load Freight North East. It was always a subsidiary of BR. The company was based in Doncaster and its main business was concentrated in the north-east of England.

Chapter 6

1   Every depot would have an establishment figure of the number of drivers required to cover all the duties, rest days and holidays to run a full service. Railways have often relied on overtime to cover the full service requirement, as shortages can be caused by sickness, accident or new workers in training. In the past both management and workforce came to rely on overtime as a way of capping the wage bill, for the former, or increasing wages, for the latter.

2   A 'Form One' is a written warning, the first formal stage in the disciplinary procedure. The new entrants in the industry may well have changed their systems, but the language inherited from BR lives on after privatisation.

3   One LUL worker interviewed, who was aware of my railway background, actually said to me that had I still been on the job, I would have been considered a dinosaur by management. At the time I was 27. Both younger and older workers talked of privatisation, or the Company Plan in the case of LUL staff, as representing a watershed in both experience and expectation. What was interesting about a large group of workers ranging in age from their mid-thirties to late forties was that their only work experience was with the railways, which would not be valued on the general labour market, and yet they were too young to take voluntary redundancy. This group was therefore in a highly ambiguous position.

References

Books and journals

Aaronovitch, S., Smith, R., Gardiner, J. and Moore, R. (1981) *The Political Economy of British Capitalism: A Marxist Analysis*. London: McGraw-Hill.

Albrow, M. (1997) *Do Organizations have Feelings?* London: Routledge.

Alcock, G.W. (1922) *Fifty Years of Railway Trade Unionism*. London: Co-operative Printing Society.

Aronowitz, S. and Cutler, J. (eds) (1998) *Post-Work: The Wages of Cybernation*. London: Routledge.

Aronowitz, S. and DiFazio, W. (1994) *The Jobless Future: Sci-Tech and the Dogma of Work*. Minneapolis: University of Minneapolis.

Bagwell, P.S. (1963) *The Railwaymen: The History of the National Union of Railwaymen, Vol. 1*. London: George Allen and Unwin.

Bagwell, P.S. (1974) *The Transport Revolution from 1770*. London: Batsford.

Bagwell, P.S. (1982) *The Railwaymen: The History of the National Union of Railwaymen, Vol. 2: The Beeching Era and After*. London: George Allen and Unwin.

Bagwell, P.S. (1984) *End of the Line?: The Fate of British Railways under Thatcher*. London: Verso.

Bagwell, P.S. (1996) *The Transport Crisis in Britain*. Nottingham: Spokesman.

Baldwin, S. (1937) *On England and other Addresses*. London: Penguin.

Barker, T.C. and Robbins, M. (1974) *A History of London Transport: Passenger Travel and the Development of the Metropolis, Vol. 2*. London: George Allen and Unwin.

Barr, C. (1993) *Ealing Studios*. 2nd edn. London: Studio Vista.

Bate, P. (1990) 'Using the culture concept in an organization development setting', *Journal of Applied Behavioural Science*, 26:1, pp. 83–106.

Bate, P. (1995) *Strategies for Culture Change*. Oxford: Butterworth-Heinemann.

Bauman, Z. (1998) *Work, Consumerism and the New Poor*. Buckingham: Open University Press.

Beck, U. (2000) *The Brave New World of Work*. Cambridge: Polity.

Behrend, G. (1969) *Gone with Regret: Recollections of the Great Western Railway 1922–1947*. 3rd edn. London: Neville Spearman.

Bell, R. (1951) *Twenty-five Years of the North Eastern Railway 1898–1922*. London: The Railway Gazette.

Betjeman, J. (1972) *London's Historic Railway Stations*. London: John Murray.

Bonavia, M.R. (1971) *The Organisation of British Railways*. London: Ian Allan.

Bonavia, M.R. (1979) *The Birth of British Rail*. London: George Allen and Unwin.

Bonavia, M.R. (1981) *British Rail: The First 25 Years*. Newton Abbot: David and Charles.

Bonavia, M.R. (1985) *Twilight of British Rail?* Newton Abbot: David and Charles.

Bradshaw, R. (1993) *Railway Lines and Levers*. Paddock Wood: Unicorn Books.

Brooke, D. (1983) *The Railway Navvy: That Despicable Race of Men*. Newton Abbot: David and Charles.

Brown, R.K. (1977) 'The growth of industrial bureaucracy – chance, choice or necessity?', in P.R. Gleichmann, J. Goudsblom and H. Korte (eds) *Human Figurations: Essays for Norbert Elias*. Amsterdam: Amsterdam Sociologisch Tijdschrift.

Brown, R.K. (1988) 'The employment relationship in sociological theory', in D. Gallie (ed.) *Employment in Britain*. Oxford: Blackwell.

Brown, R.K. (1992) *Understanding Industrial Organisations: Theoretical Perspectives in Industrial Sociology*. London: Routledge.

Brown, R.K. (1993) 'The negotiated order of the industrial enterprise', in H. Martins (ed.) *Knowledge and Passion: Essays in Honour of John Rex*. London: I.B. Traus.

Bulmer, M. (ed.) (1975) *Working-class Images of Society*. London: Routledge and Kegan Paul.

Cain, P.J. (1988) 'Railways 1870–1914: the Maturity of the Private System', in M.J. Freeman and D.H. Aldcroft (eds) *Transport in Victorian Britain*. Manchester: Manchester University Press.

Carey, J. (1992) *The Intellectuals and the Masses: Pride and Prejudice among the Literary Intelligentsia, 1880–1939*. London: Faber and Faber.

Carey, J. (ed.) (1999) *The Faber Book of Utopias*. London: Faber.

Carroll, L. (1998) [1865] *Alice's Adventures in Wonderland*. London: Penguin.

Casey, C. (1995) *Work, Self and Society after Industrialism*. London: Routledge.

Castens, S. (2000) *On the Trail of the Titfield Thunderbolt*. Bath: Thunderbolt Books.

Champy, J. (1995) *Reengineering Management: The Mandate for New Leadership*. London: Harper Collins.

Coats, A.W. (ed.) (1971) *The Classical Economists and Economic Policy*. London: Methuen.

Coleman, T. (1986) *The Railway Navvies: A History of the Men who Made the Railways*. London: Penguin.

Cottrell, W.F. (1951) 'Death by dieselization: a case study in the reaction to technological change', *American Sociological Review*, 16, pp. 358–65.

Crompton, G. (1995) 'The railway companies and the nationalisation issue 1920–50', in R. Millward and J. Singleton (eds) *The Political Economy of Nationalisation in Britain 1920–50*. Cambridge: Cambridge University Press.

Crouzet, F. (1982) *The Victorian Economy*. London: Methuen.

Cunningham, M. (1993) '"From the ground up"?: The Labour governments and economic planning', in J. Fyrth (ed.) *Labour's High Noon: The Government and the Economy 1945–51*. London: Lawrence and Wishart.

Davis, F. (1979) *Yearning for Yesterday: A Sociology of Nostalgia*. New York: Free Press.

Deal, T. and Kennedy, A. (1988) *Corporate Cultures: The Rites and Rituals of Corporate Life*. London: Penguin.

Department of Transport (1994) *Britain's Railways: A New Era*. London: The Department of Transport.

Department of Transport (1995) *New Safeguards for Rail Services*. London: The Department of Transport.

Dixon, R. and Muthesius, S. (1985) *Victorian Architecture*. 2nd edn. London: Thames and Hudson.

du Gay, P. (1996) *Consumption and Identity at Work*. London: Sage.

du Gay, P. and Salaman, G. (1992) 'The cult(ure) of the customer', *Journal of Management Studies*, 29:4, pp. 616–33.

Dyos, H.J. and Aldcroft, D.H. (1974) *British Transport: An Economic Survey from the Seventeenth Century to the Twentieth*. London: Penguin.

Eagleton, T. (2000) *The Idea of Culture*. Oxford: Blackwell.

Edmonds, J. (2000) 'Creating Railtrack', in R. Freeman and J. Shaw (eds) *All Change: British Railway Privatisation*. London: McGraw-Hill.

Edwards, P.K. and Whitston, C. (1994) 'Disciplinary practice: a study of railways in Britain, 1860–1988', *Work, Employment and Society*, 8:3, pp. 317–37.

Edwards, R. (1979) *Contested Terrain: The Transformation of the Workplace in the Twentieth Century*. London: Heinemann.

Eldridge, J.E.T. (1968) *Industrial Disputes; Essays in the Sociology of Industrial Relations*. London: Routledge and Kegan Paul.

Eldridge, J.E.T. (ed.) (1970) *Max Weber: The Interpretation of Social Reality*. London: Nelson.

Elliot, J. (1982) *On and Off the Rails*. London: George Allen and Unwin.

Faludi, S. (1999) *Stiffed: The Betrayal of Modern Man*. London: Vintage.

Farrington, J. (1984) *Life on the Lines*. Ashbourne: Moorland Publishing.

Ferner, A. (1988) *Governments, Managers and Industrial Relations: Public Enterprises and their Political Environment*. Oxford: Basil Blackwell.

Ferner, A. (1989) *Ten Years of Thatcherism: Changing Industrial Relations in British Public Enterprises*. Warwick Papers in Industrial Relations, 27.

Ferneyhough, F. (1983) *Steam Up! A Railwayman Remembers*. London: Robert Hale.

Fiennes, G.F. (1973) *I Tried to Run a Railway*. Shepperton: Ian Allan.

Fineman, S. (ed.) (1993) *Emotion in Organisations*. London: Sage.

Foreman-Peck, J. and Millward, R. (1994) *Public and Private Ownership of British Industry 1820–1990*. Oxford: Oxford University Press.

Freeman, R. and Shaw, J. (eds) (2000) *All Change: British Railway Privatisation*. London: McGraw-Hill.

Friend, H. (1994a) *Track Record*. Durham: Aldan, MacNair and Young.

Friend, H. (1994b) 'Driver training on the class 40s', *Back Track*, 8:5, pp. 241–8.

Furedi, F. (1992) *Mythical Past, Elusive Future: History and Society in an Anxious Age*. London: Pluto.

Gabriel, Y. (1993) 'Organizational nostalgia – reflections on "The Golden Age"', in S. Fineman (ed.) *Emotion in Organisations*. London: Sage.

Garrahan, P. and Stewart, P. (1992) *The Nissan Enigma: Flexibility at Work in a Local Economy*. London: Mansell.

Gasson, H. (1981) *Signalling Days: Final Reminiscences of a Great Western Railwayman*. Oxford: Oxford Publishing Company.

Glaister, S. and Travers, T. (1993) *New Directions for British Railways? The Political Economy of Privatisation and Regulation*. London: Institute of Economic Affairs.

Goldthorpe, J.H., Lockwood, D., Bechhofer, F. and Platt, J. (1968) *The Affluent Worker: Industrial Attitudes and Behaviour*. Cambridge: Cambridge University Press.

Goodrich, C.L. (1975) *The Frontier of Control: A Study in British Workshop Politics*. London: Pluto Press.

Gorz, A. (1999) *Reclaiming Work: Beyond the Wage-based Society*. Cambridge: Polity.

Gouldner, A.W. (1964) *Patterns of Industrial Bureaucracy*. New York: Free Press.

Gourvish, T.R. (1972) *Mark Huish and the London North Western Railway: a Study of Management*. Leicester: Leicester University Press.

Gourvish, T.R. (1986) *British Railways 1948–73: A Business History*. Cambridge: Cambridge University Press.

Gourvish, T.R. (1990) 'British Rail's "business-led" organization, 1977–1990: government-industry relations in Britain's public sector', *Business History Review* 64:Spring, pp. 109–49.

Graham, L. (1995) *On the Line at Subaru-Isuzu: The Japanese Model and the American Worker*. Ithaca: ILR Press/Cornell University Press.

Gratton, B. (1990) '"A triumph of modern philanthropy": age criteria in labor management at the Pennsylvania Railroad, 1875–1930', *Business History Review*, 64:Winter, pp. 630–56.

Grey, C. (1994) 'Career as a projection of the self and labour process discipline', *Sociology*, 28:2, pp. 479–97.

Grint, K. (1995) *Management: A Sociological Introduction*. Cambridge: Polity.

Groome, C. (1986) *The Decline and Fall of the Engine Driver*. London: Groome.

Guest, D.E., Peccei, R. and Fulcher, A. (1993) 'Culture change and quality improvement in British Rail', in D. Gowler, K. Legge and C. Clegg (eds) *Case Studies in Organisational Behaviour and Human Resource Management*. London: Paul Chapman Publishing.

Hall, S. and Jacques, M. (eds) (1987) *The Politics of Thatcherism*. London: Lawrence and Wishart.

Hardy, R.H.N. (1989) *Beeching: Champion of the Railway?* London: Ian Allan.

Haresnape, B. (1979) *British Rail 1948–78: A Journey by Design*. Shepperton: Ian Allan.

Haresnape, B. (1981) *British Rail Fleet Survey, 1: Early Prototype and Pilot Scheme Diesel-Electrics*. London: Ian Allan.

Harris, N. (1997) *Railway Paintings by Cuneo: A Retrospective*. Peterborough: Emap Apex Publications.

Harris, N.G. and Godward, E. (1997) *The Privatisation of British Rail*. London: The Railway Consultancy Press.

Hartley, L.P. (1958) *The Go-Between*. London: Penguin.

Hay, C. (1996) 'Narrating crisis: the discursive construction of the "Winter of Discontent"', *Sociology*, 30:2, pp. 253–77.

Henshaw, D. (1991) *The Great Railway Conspiracy*. Hawes: Leading Edge.

Hewison, R. (1987) *The Heritage Industry: Britain in a Climate of Decline*. London: Methuen.

Hick, F.L. (1991) *That Was My Railway: From Ploughman's Kid to Railway Boss, 1922–1969*. Kettering: Silver Link Publishing.

HMSO (1963) *The Reshaping of the British Railways: Parts 1 and 2*. London: HMSO.

Hobsbawm, E. (1986) *Labouring Men: Studies in the History of Labour*. London: Weidenfeld and Nicolson.

Hobsbawm, E. and Ranger, T. (eds) (1992) *The Invention of Tradition*. Cambridge: Cambridge University Press.

Hoggart, R. (1957) *The uses of literacy: Aspects of working-class life with special reference to publications and entertainments*. London: Chatto and Windus.

Hollowell, P. (1975) 'The remarkably constant fortunes of the British Railways locomotiveman', in G. Esland, G. Salaman and M. Speakman (eds) *People and Work*. Edinburgh: Open University Press.

Howell, D. (1999) *Respectable Radicals: Studies in the Politics of Railway Trade Unionism*. Aldershot: Ashgate.

Hudson, M. (1995) *Coming Back Brokens: A Year in a Mining Village*. London: Vintage.

Irving, R.J. (1976) *The North Eastern Railway Company 1870–1914: An Economic History*. Leicester: Leicester University Press.

Irving, R.J. (1978) 'The profitability and performance of British Railways, 1870–1914', *Economic History Review*, 31, pp. 46–66.

Jack, I. (2001) *The Crash that Stopped Britain*. London: Granta.

Joy, S. (1973) *The Train that Ran Away: The Inside Story of British Railways' Chronic Financial Failures since Nationalisation*. London: Ian Allan.

Kellett, J.R. (1969) 'Writing on Victorian railways: an essay in nostalgia', *Victorian Studies*, 13:1, pp. 90–6.

Kingsford, P.W. (1970) *Victorian Railwaymen: The Emergence and Growth of Railway Labour 1830–1870*. London: Frank Cass.

Le Goff, J. (1992) *History and Memory*. New York: Columbia University Press.

Leys, C. (1989) *Politics in Britain: From Labourism to Thatcherism*. London: Verso.

Lockwood, D. (1975) 'Sources in variation in working-class images of society', in M. Bulmer (ed.) *Working-class Images of Society*. London: Routledge and Kegan Paul.

Lowenthal, D. (1998) *The Heritage Crusade and the Spoils of History*. Cambridge: Cambridge University Press.

Lynn-Meek, V. (1988) 'Organizational culture: origins and weaknesses', *Organization Studies*, 9:4, pp. 453–73.

McCord, N. (1979) *North East England: The Region's Development 1760–1960*. London: Batsford Academic.

McKenna, F. (1976) 'Victorian railway workers', *History Workshop Journal*, 1:Spring, pp. 26–73.

McKenna, F. (1980) *The Railway Workers 1840–1970*. London: Faber and Faber.

McKibbin, R. (1991) 'Homage to Wilson and Callaghan', *London Review of Books*, 13:20, pp. 3–5.

MacInnes, J. (1987) *Thatcherism at Work: Industrial Relations and Economic Change*. Milton Keynes: Open University Press.

McKillop, N. (1950) *The Lighted Flame: A History of the Associated Society of Locomotive Engineers and Firemen*. London: Nelson.

Marris, P. (1974) *Loss and change*. London: Routledge and Kegan Paul.

Marx, K. (1954) *Capital, Vol. 1*. London: Lawrence and Wishart.

Marx, K. (1976) [1867] *Capital, Vol. 1*. Translated by B. Fowkes. London: Penguin.

Mason, F. (1992) *Life Adventure in Steam: A Merseyside Driver Remembers*. Birkenhead: Countyvise.

Mead, M. (1970) *Culture and Commitment: The New Relationships between Generations in the 1970s*. New York: Doubleday.

Mead, M. (1978) *Culture and Commitment: The New Relationships Between Generations in the 1970s*. New York: Anchor Press/ Doubleday.

Meakin, D. (1976) *Man and Work: Literature and Culture in Industrial Society*. London: Methuen.

Metzgar, J. (2000) *Striking Steel: Solidarity Remembered*. Philadelphia: Temple University Press.

Milkman, R. (1997) *Farewell to the Factory*. Berkeley: University of California Press.

Millward, R. and Singleton, J. (eds) (1995) *The Political Economy of Nationalisation in Britain 1920–50*. Cambridge: Cambridge University Press.

Milne, A.M. and Laing, A. (1956) *The Obligation to Carry: An Examination of the Development of the Obligations Owed to the Public by British Railways*. London: Institute of Transport.

Munro, R. (1998) 'Belonging on the move: market rhetoric and the future as obligatory passage', *The Sociological Review*, 46:2, pp. 208–43.

Murphy, B. (1980) *ASLEF 1880–1980: A Hundred Years of the Locoman's Trade Union*. London: ASLEF.

Newbould, D.A. (1985) *Yesterday's Railwayman*. Poole: Oxford Publishing Company.

Nichols, T. (1986) *The British Worker Question: A New Look at Workers and Productivity in Manufacturing*. London: Routledge and Kegan Paul.

Nichols, T. and Beynon, H. (1977) *Living with Capitalism: Class Relations and the Modern Factory*. London: Routledge and Kegan Paul.

North, G.A. (1975) *Teesside's Economic Heritage*. Middlesborough: County Council of Cleveland.

O'Connell Davidson, J. (1993) *Privatization and Employment Relations: The Case of the Water Industry*. London: Mansell.

Orwell, G. (1984) *The Penguin Essays of George Orwell*. London: Penguin.

Outram, Q. (1997) 'The stupidest men in England? The industrial relations strategy of the coalowners between the lockouts, 1923–1924', *Historical Studies in Industrial Relations*, 4, pp. 65–95.

Parker, M. (2000) *Organizational Culture and Identity: Unity and Division at Work*. London: Sage.

Parker, P. (1991) *For Starters: The Business of Life*. London: Pan.

Payton, P. (1997) 'An English cross-country railway: rural England and the cultural reconstruction of the Somerset and Dorset Railway', Unpublished paper presented to the Railway Identity and Place Conference, Truro.

Pearson, A.J. (1967) *Man of the Rail*. London: George Allen and Unwin.

Pelling, H. (1976) *A Short History of the Labour Party*. 5th edn. London: Macmillan.

Pendleton, A. (1984) 'Managerial strategy, new technology and industrial relations: the case of railway signalling', Unpublished paper presented to the Labour Process Conference, Manchester.

Pendleton, A. (1991a) 'The barriers to flexibility: flexible rostering on the railways', *Work, Employment and Society*, 5:2, pp. 241–57.

Pendleton, A. (1991b) 'Integration and dealignment in public enterprise industrial relations: a study of British Rail', *British Journal of Industrial Relations*, 29:3, pp. 411–26.

Pendleton, A. (1993) 'Rail', in A. Pendleton and J. Winterton (eds) *Public Enterprise in Transition: Industrial Relations in State and Privatized Industries*. London: Routledge.

Pendleton, A. (1995) 'The emergence and use of quality in British Rail', in I. Kirkpatrick and M. Martinez Lucio (eds) *The Politics of Quality in the Public Sector: The Management of Change*. London: Routledge.

Pendleton, A. and Winterton, J. (eds) (1993) *Public Enterprise in Transition: Industrial Relations in State and Privatized Industries*. London: Routledge.

Penn, R. (1986) 'Socialisation into skilled identities', Unpublished paper presented to Labour Process Conference, Aston University.

Penn, R. (1994) 'Technical change and skilled manual work in contemporary Rochdale', in R. Penn, M. Rose, M. and J. Rubery (eds) *Skill and Occupational Change*. Oxford: Oxford University Press.

Perry, G. (1981) *Forever Ealing: A Celebration of the Great British Film Studio*. London: Pavilion.

Peters, T. (1992) *Liberation Management: Necessary Disorganization for the Nanosecond Nineties*. London: Macmillan.

Peters, T. (1995) *The Pursuit of Wow! Every Person's Guide to Topsy-Turvy Times*. London: Macmillan.

Peters, T. and Austin, N. (1986) *A Passion for Excellence: The Leadership Difference*. London: Fontana.

Peters, T. and Waterman, R.H. (1982) *In Search of Excellence: Lessons from America's Best-run Companies*. London: Harper and Row.

Pollins, H. (1971) *Britain's Railways: An Industrial History*. Newton Abbot: David and Charles.

Pollitt, C. (1993) *Managerialism and the Public Services: Cuts or Culture Change in the 1990s?* 2nd edn. Oxford: Blackwell.

Potts, C.R. (1996) *The GWR and the General Strike*. Oxford: Oakwood.

Powell, D. (1993) *The Power Game: The Struggle for Coal*. London: Duckworth.

Putnam, L. and Mumby, D. (1993) 'Organizations, emotion and the myth of rationality', in S. Fineman (ed.) *Emotion in Organisations*. London: Sage.

Railtrack (1997) *1996/97 Annual Report and Accounts*. London: Railtrack.

Railtrack (1998) *1997/98 Annual Report and Accounts*. London: Railtrack.

Ramsay, H. (1996) 'Managing sceptically: a critique of organizational fashion', in S.R. Clegg and G. Palmer (eds) *The Politics of Management Knowledge*. London: Sage.

Raynes, J.R. (1921) *Engines and Men: The History of the Associated Society of Locomotive Engineers and Firemen*. Leeds: Goodall and Suddick.

Rees, R. (1994) 'Economic aspects of privatization in Britain', in V. Wright (ed.) *Privatization in Western Europe: Pressures, Problems and Paradoxes*. London: Pinter.

Revill, G.E. (1989) 'Paternalism, community and corporate culture: a study of the Derby headquarters of the Midland Railway Company and its workforce 1840–1900', Unpublished PhD thesis, Loughborough University of Technology.

Revill, G. (1991) 'Trained for life: personal identity and the meaning of work in the nineteenth-century railway industry', in C. Philo (ed.) *New Words, New Worlds: Reconceptualising Social and Cultural Geography*. Lampeter: Department of Geography, St. David's University College.

Reynolds, M. (1968) [1881] *Engine Driving Life: Stirring Adventures and Incidents in the Lives of Locomotive Engine-drivers*. London: Hugh Evelyn.

Richards, J. and MacKenzie, J.M. (1988) *The Railway Station: A Social History*. Oxford: Oxford University Press.

Rifkin, J. (1996) *The End of Work: The Decline of the Global Labor Force and the Dawn of the Post-market Era*. New York: Putnam.

Roberts, I. (1993) *Craft, Class and Control: The Sociology of a Shipbuilding Community*. Edinburgh: Edinburgh University Press.

Roberts, I. (1999) 'A historical construction of the working class', in H. Beynon and P. Glavanis (eds) *Patterns of Social Inequality: Essays for Richard Brown.* Harlow: Longman.

Rolt, L.T.C. (1961) *Railway Adventure.* London: Pan.

Rolt, L.T.C. (1978) *Narrow Boat.* London: Methuen.

Rowbotham, S. (2000) *The Promise of a Dream: Remembering the Sixties.* London: Penguin.

Salaman, G. (1974) *Community and Occupation: An Exploration of Work/Leisure Relationships.* Cambridge: Cambridge University Press.

Salaman, J.G. (1969) 'Some sociological determinants of occupational communities', Unpublished PhD thesis, Oxford University.

Samuel, R. (1994) *Theatres of Memory, Vol. 1: Past and Present in Contemporary Culture.* London: Verso.

Samuel, R. (1998) *Theatres of Memory, Vol. 2: Island Stories: Unravelling Britain.* London: Verso.

Savage, M. (1998) 'Discipline, surveillance and the "career": Employment on the Great Western Railway 1833–1914', in A. McKinlay and K. Starkey (eds) *Foucault, Management and Organization Theory.* London: Sage.

Saville, R. (1993) 'Commanding heights: the nationalisation programme', in J. Fyrth (ed.) *Labour's High Noon: The Government and the Economy 1945–51.* London: Lawrence and Wishart.

Seabrook, J. (1978) *What Went Wrong?* London: Victor Gollancz.

Seabrook, J. (1982) *Unemployment.* London: Quartet Books.

Seabrook, J. (1988) *The Race for Riches.* London: Marshall Pickering.

Seldon, A. (1998) *Major: A Political Life.* London: Phoenix.

Semmens, P.W.B. (1966) *Bill Hoole: Engineman Extraordinary.* London: Ian Allan.

Sennett, R. (1996) *The Uses of Disorder: Personal Identity and City Life.* London: Faber.

Sennett, R. (1998) *The Corrosion of Character: The Personal Consequences of Work in the New Capitalism.* London: Norton.

Shinwell, E. (1955) *Conflict without Malice.* London: Oldhams Press.

Simmons, J. (1994) *The Express Train and other Railway Studies.* Nairn: David St John Thomas.

Smith, K. (1989) *The British Economic Crisis: Its Past and Future.* London: Penguin.

Smith, M. (2000) *Culture: Reinventing the Social Sciences.* Buckingham: Open University Press.

Smith, P.W. (1972) *Mendips Engineman.* Oxford: Oxford Publishing Company.

Sosteric, M. (1996) 'Subjectivity and the labour process: a case study in the restaurant industry', *Work, Employment and Society*, 10:2, pp. 297–318.

Stafford, W. (1989) '"This once happy country": nostalgia for pre-modern society', in C. Shaw and M. Chase (eds) *The Imagined Past: History and Nostalgia.* Manchester: Manchester University Press.

Stedman Jones, G. (1992) *Outcast London: A Study in the Relationship between Classes in Victorian Society.* London: Penguin.

Steiger, T.L. (1993) 'Construction skill and skill construction', *Work Employment and Society*, 7:4, pp. 535–60.

Stein, M.B. (1978) 'The meaning of skill: the case of the French engine-drivers, 1837–1917', *Politics and Society*, 8:3–4, pp. 399–427.

Stewart, B. (1982) *On the Right Lines: Footplate Memories.* Woodchester: Peter Watts.

Storey, J. (1992) *Developments in the Management of Human Resources: An Analytical Review.* Oxford: Blackwell.

Storey, J. and Sisson, K. (1993) *Managing Human Resources and Industrial Relations.* Buckingham: Open University Press.

Strangleman, T. (1998) 'Railway and grade: the historical construction of contemporary identities', Unpublished PhD thesis, University of Durham.

Strangleman, T. (2002) 'Constructing the past: railway history from below or a study in nostalgia?', *The Journal of Transport History*, 23:2, pp. 147–58.

Strangleman, T. and Roberts, I. (1999) 'Looking through the window of opportunity: the cultural cleansing of workplace identity', *Sociology*, 33:1, pp. 47–67.

Swann, P. (1989) *The British Documentary Film Movement 1926–1946.* Cambridge: Cambridge University Press.

Sykes, R., Austin, A., Fuller, M., Kinoshita, T. and Shrimpton, A. (1997) 'Steam attraction: railways in Britain's national heritage', *The Journal of Transport History*, 3:2, pp. 156–75.

Taylor, A.J. (1972) *Laissez-faire and State Intervention in Nineteenth-century Britain.* London: Macmillan.

Terkel, S. (1990) *Working: People Talk About What They Do All Day and How They Feel About What They Do.* New York: Pantheon Books.

Thompson, A.W.J. and Hunter, L.C. (1973) *The Nationalised Transport Industries.* London: Heinemann.

Tomlinson, J. (1982) *The Unequal Struggle? British Socialism and the Capitalist Enterprise.* London: Methuen.

Tomlinson, J. (1997) *Democratic Socialism and Economic Policy: The Attlee Years 1945–1951.* Cambridge: Cambridge University Press.

Turnbull, P. and Wass, V. (1995) 'Anorexic organisations and labour market lemons: the (mis)management of redundancy in Britain', Unpublished paper presented to the British Universities Industrial Relations Association Annual Conference, Durham.

Turner, A. (1996) *Careering with Steam.* Bath: Millstream Books.

Turner, B.S. (1987) 'A note on nostalgia', *Theory, Culture and Society*, 4:1, pp. 147–56.

Uff, J. (2000) *The Southall Rail Accident Inquiry Report.* Norwich: HSE Books.

Vaughan, A. (1984) *Signalman's Morning/Signalman's Twilight.* London: Pan.

Vaughan, A. (1987) *Signalman's nightmare.* London: Guild Publishing.

Vincent, M. and Green, C. (eds) (1994) *The Intercity Story.* Sparkford: Oxford Publishing Company.

Weber, M. (1964) *The Theory of Social and Economic Organizations.* New York: The Free Press.

Wedderburn, D. (1965) *Redundancy and the Railwaymen.* Cambridge: Cambridge University Press.

Whittaker, N. (1995) *Platform Souls: The Trainspotter as Twentieth-century Hero.* London: Victor Gollancz.

Williams, K., Haslam, C., Johal, S., Froud, J., Shaoul, J. and Williams, J. (1996) *The Right Argument: refocusing the Debate on Privatization and Marketization.* Manchester International Centre for Labour Studies Working Paper 11, Manchester.

Williams, R. (1973) *The Country and the City.* Oxford: Oxford University Press.

Williams, R. (1983) *Keywords: A Vocabulary of Culture and Society*. London: Fontana.

Willmott, H. (1993) 'Strength is ignorance; slavery is freedom: managing culture in modern organizations', *Journal of Management Studies*, 30:4, pp. 515–52.

Wilson, C.D. (1996) *Forward!: The Revolution in the Lives of the Footplatemen 1962–1996*. Far Thrupp: Sutton Publishing.

Wolmar, C. (1997) 'Can Branson get back on track?', *New Statesman*, 7 November.

Wolmar, C. (2000) 'Creating the passenger rail franchises', in R. Freeman and J. Shaw (eds) *All Change: British Railway Privatisation*. London: McGraw-Hill.

Wolmar, C. and Ford, R. (2000) 'Selling the passenger railway', in R. Freeman and J. Shaw (eds) *All Change: British Railway Privatisation*. London: McGraw-Hill.

Wright, P. (1985) *On Living in an Old Country: The National Past in Contemporary Britain*. London: Verso.

Wright, P. (1993) *A Journey through Ruins: A Keyhole Portrait of British Postwar Life and Culture*. London: Flamingo.

Wright, P. (1996) *The Village that Died for England: The Strange Story of Tyneham*. London: Vintage.

Wright, S. (ed.) (1994) *Anthropology of Organizations*. London: Routledge.

Wright Mills, C. (1967) *The Sociological Imagination*. Oxford: Oxford University Press.

Wright, V. (ed.) (1994) *Privatization in Western Europe: Pressures, Problems and Paradoxes*. London: Pinter.

Young, M. and Willmott, P. (1957) *Family and Kinship in East London*. London: Routledge and Kegan Paul.

Zweig, F. (1952) *The British Worker*. London: Penguin.

General circulation newspapers and magazines

Back Track
Financial Times
Guardian
Independent
Independent on Sunday
London Review of Books
New Statesman
Punch
Rail
The Railway Magazine

In-house and trade union newspapers and magazines

Hotline (Virgin Trains)
Livewire (GNER/ ICEC)
Loadhaul News: The Newspaper of Loadhaul
Locomotive Journal (ASLEF)
North East News: The Newspaper of Trainload Freight North East
North Eastern Railway Magazine
U Magazine: The Magazine for London Underground People

Index

ABN Amro 169
accidents 14, 24, 36, 95, 171
    see also Castlecary; Hatfield;
        Ladbroke Grove; Southall
adaptability 161
adolescence 138, 159
    engineered 143, 176
advancement 21, 23, 24
affinity 28
age 97, 98, 141, 143, 148
air traffic control 174
Aldcroft, D.H. 16, 64
Allen, W.P. 45
Alston 81
amalgamations 18, 28, 92, 104
Anstey, Edgar 101
anxiety 155, 156
apprenticeships 29, 49, 53–4, 118
Armstrong, Geoff 160–1
Aronowitz, S. 10
ASLEF (Associated Society of
    Locomotive Engineers and
    Firemen) 32, 36–9, 45, 69, 111,
    112, 147, 148, 156
ASRS (Amalgamated Society of
    Railway Servants) 36, 37, 39
assets 45, 48, 122
Associated British Ports 5
attachment 102, 106, 109, 136
    emotional 138
    sentimental 107, 110
Attlee, Clement 3, 44, 47, 48, 174
Austin, N. 152
autonomy 28, 41, 79, 154, 159
    regional 78

Bagwell, P.S. 27, 36, 64, 74, 93, 111,
    118, 120, 121, 154
bailiwick 28
Baldwin, Stanley 118
Balfour Beatty 165
bankruptcies 16
bar boys 29

Barnes, Alfred 47
Barr, C. 85, 118
barriers to entry 17
Barrington-Ward, Michael 45
Bass, Michael Thomas 36
Bate, P. 114, 115
Bath 29
Baty (general secretary, ASLEF) 38–9
Bauman, Z. 10
Beck, U. 10–11
Beeching, Richard 44, 55, 56, 57–8,
    59, 62, 63, 64, 69, 80–5, 91
Behrend, George 87–8, 98
Benstead, John 46
Betjeman, John 55–6, 60
Beynon, H. 109
Big Four era 118
Blair, Tony 173, 174
blue-collar workers 23, 72, 90, 171
    divisions between white-collar and
        144
boilerwashers 74
Bonavia, M. 22, 43, 44, 62, 63, 67,
    69, 94
bonus payments 147
booking clerks 84
booking lad role 33
box boys 96
BP (British Petroleum) 5
BR (British Railways) 1, 3, 12, 19,
    44, 55, 56, 58, 63, 71, 74, 115
    architects 60
    comradeship 66
    continuity in senior personnel 48
    gradual withdrawal from particular
        types of traffic 80
    less selective in recruitment 94
    sale of profitable parts of empire
        113
    senior managers 50, 121
    steam locomotives ended on
        mainlines 90
    support from public funds for 111

Bradshaw, R.   33, 34, 71, 84, 92, 93, 95, 96
branch lines   81
Branson, Richard   130–1, 151
Braybrook, Ian   125–6
BRB (British Railways Board)   44, 57, 61, 64, 94, 115
  see also Parker (Peter); Reid; Welsby
bridges   17
'British disease'   7
British Gas   5, 6
Brown, Joe   31
Brown, J.G.   23
Brown, R.K.   15, 91, 143
brownfield sites   136, 159–61
Brunel, I.K.   61
BT (British Telecom)   5, 6, 122
BTC (British Transport Commission)   3, 19, 44, 46, 48, 54, 71
  abolished   57
  Modernisation Plan (1955)   54–5, 74, 76, 80, 111
BTC plans to restructure   118
Bulmer, M.   91
bureaucracy   65, 73, 114
  attack on   8
  centralised   15
  imagination and creativity held back by inertia of   104
  resistance to   85
  unnecessary layers of   54
Butler, R.A.   5
Byers, Stephen   170

Cable and Wireless   5
Cain, P.J.   16
Callaghan administration   48
Cambrian Railways   87
capital   9, 15, 17, 86, 99
  borrowing   48
  crisis of   135
  direct clash between labour and   50
capital intensity   17
capital stock   3, 16
capitalism   11, 19, 100, 160
  'new'   11, 175
car industry   3, 92–3
career trajectories   22–3

Carey, John   47
Casey, C.   10, 109, 136, 158
Castens, S.   85
Castle, Barbara   63
Castlecary   34
casual workers   26
CBI (Confederation of British Industry)   160
central economic planning   45, 47
centralisation   41, 54
Challow   82
Champy, J.   152, 160
change   67, 85, 99
  arbitrarily imposed   176
  bitterness over   84
  coping with   102, 161
  major   23
  necessary in order to return to utopian past   104–5
  power to resist or block   136
  remembered through the personal   81
  social relations threatened by   82
  will to enact   87
  see also culture change
Charing Cross   111
Chile   26
Clark, Alan   111
class war   86, 112
cleaners   1, 23, 27, 29, 71, 72, 74
Clementson, Tom (Tiny)   71
closures   55, 69, 80–5, 91, 98
  impending   93
  mass   19
coal   3, 5
  miners   101
  recession in   18
  shipped   26
coalmen   74
coats of arms   41
Coats, A.W.   16
Coggett (National Association of Railway Workers)   86
Coleman, T.   19
commercialisation   105, 126, 154, 172, 174
commitment to work   137, 151, 152, 176
  identification and   103, 154–8

commitment to work – *continued*
  lack of   86, 157, 158
  long-term   84
community   81
Community of European Railways
  111
companionship   89
Company Plan   116
compensation
  seniority-based   24
  shareholder   48, 170
competition   17, 18, 122
  desire to create as much as possible
    153
  intense   16
  new source of   18
  road   19, 57
competitive pressures   56
concentration   16, 17, 18
conservatism   108, 176
  defence of   161
  organisational   117
  perceived   102, 124
Conservative Party/Governments   4,
  5, 6–7, 11, 54, 66, 110, 113, 174
  libertarian free-marketeers and
    social conservatives   172
  neo-liberals   5, 104
Consett   32
construction   19
contingency   9
contractors   148, 160
  poorly trained   151
  squeezing   149
control
  frontier of   135, 142
  Government/State   18, 174
  management   15
  new structures   15
  public   3, 7
  technologies of   9, 10
  worker   136
Cooper Brothers   64
Corbett, Gerald   167
Cottrell, W.F.   74
Coucher, Iain   177
Country Life   87
covenants   25
Cox, Ted   33

cranemen   74
crashes *see* accidents
credit rating   169
Crompton, G.   18
Crouzet, F.   16
culture   39–42, 44, 61, 80, 161
  business-led   115
  central and vital aspect of   77
  customer-first   164
  fresh   126
  notions of   105–6
  problematic   114
  shared   34
  strong autonomous   78
  subtle shifts in   67
  understanding of   69
culture change   2, 7, 8, 9–11, 106,
  123, 126, 131, 134, 175
  attempt to promote commitment
    154
  contradictions of   152
  experiencing   139–46
  marginalised by   158
  ultimate contradiction for
    management engaged in   159
  understanding   135–9
Cuneo, Terence   131
Cunningham, M.   45, 47
customs   15, 28, 29
Cutler, J.   10

Daily Express   170
danger   1, 29
Davis, F.   102, 108, 109, 110, 119
Deal, T.   106, 152
debt   48, 170
decision-making   79
  aim to liberate and devolve   114
  blurring   5
  industrial, more democratic
    extension of   47
  lack of scope for workers'
    participation in   46
  overly bureaucratic   165
decline   85, 88, 89, 167
  connected with the arrival of
    newcomers   98
  regional aspect of is   98
  urban recruitment blamed for   95

demand for workers/labour 19, 72
demotion 88
denationalisation 117, 153
Deownly, George 73, 80, 81, 97
dependency 7, 25, 41
depersonalisation 21
depression 25
  (1930s) 18, 19, 27
deregulation 149
deskilling 74–80, 96
DFM (Documentary Film Movement)
  101
diesels 74, 76–7, 79, 96, 97
differentials in pay 148
dignity 159
discipline 25, 151
  strict 22
dismissal 24
division of labour 15, 21
Dixon, R. 60
Docks and Inland Waterways 44
Dowlais iron works 15
Driver Only Operation 124
drivers 21, 22, 26, 27, 29, 30, 31, 32,
  50, 70, 72, 93, 96, 140
  companies that have a requirement
    for 146–7
  converted to diesels 76–7
  described as being formidable
    fellows 141
  job went when the steam engine
    went 97
  junior 23
  number fallen 74
  promotion to 99
  reluctance to train 148
  retired 98, 124
  shortage of 147–8
drunkenness 19
du Gay, P. 10, 109, 136, 158
Durham 71
Durkheim, Emile 100
Dyos, H.J. 16, 64

Eagleton, T. 41
Ealing Studios 55, 85, 118
earth works 17
economies of scale 17, 18, 174
Eden, Anthony 5

Edinburgh 155
Edmonds, J. 120, 153, 169
Edwards, R. 135
efficiency 3, 18, 43, 45, 87
  belief in 6
  market pressures and signals to
    produce greater levels of 5
  national 4
  obsession with 85
  unnecessary search for 82
Eldridge, J.E.T. 73
electric motive power 54–5, 60
electricity 5, 6
electrification 71
electromechanical and electronic
  equipment 74–6
Eliot, George 100
Elliot, John 46, 47, 51–2, 66
embankments 17
emotion 61, 102, 107, 110
  direct annexation of 130
  stimulated by discomfort with the
    present 119
employment 3, 21, 26
  alienating, estranging but ultimately
    necessary forms of 100
  alternative 154
  better-paid 23
  car industry 92
  competition from other forms of
    95
  continuity of 81
  decline in 92
  deterioration in terms and
    conditions 148
  full 92, 94
  heroic 101
  inequality in the relationship 36
  loss of rights 155
  numbers 19
  outsiders 59
  respectable and permanent 19
  secure, continuous, undermined
    image of 94
  stable 25
  stereotyped 39
engineering posts 166
England 156
  Home Counties 164

England – *continued*
   north-east   18, 70, 72, 76, 80, 90,
       94, 98
   north-west   60
   rail cut off from Scotland   164
   south-east   99
   south-west   147
   West Midlands   15
English Electric   79
Euston   60
Evans, Huw   142, 157
EWS (English, Welsh and Scottish
   Railways)   12, 125, 127, 164,
   165
exploitation   86

Fabian Society   3
Fairless, Stan   80
Faludi, Susan   34
Far East   7
fault-finding   77
Ferner, A.   16, 113, 114, 154
Ferneyhough, Frank   49, 50–1, 141
Fiennes, Gerard   49–50, 52, 53, 60,
   65
Filby, Jon   155
Financial Times   166, 167, 169, 177
Fineman, S.   61
fireboxes   29, 49–50
firedroppers   74
firemen   21, 23, 27, 30–1, 32, 70, 71,
   72, 73, 98, 99
   re-labelled second men   89
fixed costs   17
flexibility   18, 116, 134, 137, 146,
   161
   barriers to   123
   linkage between youth and   142
   severely curtailed   155
Footplate Days and Ways   1, 176
footplatemen   21, 28, 30, 36, 76
Ford, R.   123
foremen   23, 26
Forth bridge   131
fragmentation   120, 121, 134, 139,
   153, 156, 165, 171, 175
   consequence of   148
   horizontal and vertical   154
   uncertainty associated with   155

work and   146–52
franchises   127, 131, 133, 155, 156
Freedland, Jonathan   169
Freeman, R.   122
freight   18, 53
   bulk high-speed   58
   reduced   84
   sector split into several parts
       119–20
   switching to the roads   64
Friend, Harry   70, 71, 76, 93–4
Furedi, F.   107–8

Gabriel, Y.   62, 106–7, 108, 109–10
Galashiels   155
Garnett, Christopher   127–8
Gasson, H.   92
Gateshead   76
GCC (Gauge Corner Cracking)   164
GDP (gross domestic product)   3, 111
General Strike (1926)   70–1
General Union of Railway Workers
   39
Gladstone, W.E.   18, 45
Glaister, S.   98, 117, 124, 155
glamour   89–90
Glancey, Jonathan   167
Glasgow   60
globalisation   9
GNER (Great North Eastern Railway)
   126, 128, 131, 164, 165
   new identity   130
   Passenger's Charter leaflet   128
GNR (Great Northern Railways
   Limited)   127–8
Godward, E.   117
Goldthorpe, J.H.   91
Goodrich, C.L.   73, 135
Gorz, A.   10, 11
Gouldner, Alvin   109–10
Gourvish, T. R.   15, 19, 45, 46, 54,
   55, 57, 58, 62, 64, 65, 74, 92, 112,
   114, 119
Government/State intervention   16,
   17, 174
grades   21, 23, 25, 29–30, 35
   blue-collar   72
   driving   22, 51, 112, 125, 148
   executive   121

footplate   34, 37, 59, 74, 78, 79, 88
  less scope to move between   155
  lower   94
  loyalty between   89
  management   44, 121, 144
  non-driving   148, 151
  reduction in number of   116
  salaried   15
  signalling   33, 34, 59, 79, 95, 98
  stability amongst   99
  supervisory   15, 84, 121
  unskilled   26
Grand Metropolitan   167
Grant, John   52
Gratton, B.   23–4, 154
Great Central Railway   56
Great Eastern Railway   104, 131
Great Western Trains   131
greedy monopolists   36
Green, Chris   120, 164, 167
greening   136, 146, 159–61
Grey, C.   24, 106
Grint, K.   139
Groome, Clive   1, 2, 89–90, 96, 98,
    99, 176
Group Standards   171
Grouping (1923)   16, 56, 174
Guardian, The   149, 166, 169
guards   36
Guild Socialism   47
GWR (Great Western Railway)   14,
    15, 17, 44, 61, 117, 118–19
  career progression in   22–3
  elegy to   87
  Swindon works   16

Hank (WR Chairman)   82
Hardy, R.H.N.   53–4, 58, 59
Hardy, Thomas   100
Haresnape, B.   60
Harris, N.   117, 131
Hartley, L.P.   104
Hatfield, Hertfordshire   164–5, 166,
    167, 169, 170
Hay, C.   6, 111
Heaton   71, 72
heavy industry   18
Henshaw, D.   59
heritage   107, 108, 131

Heseltine, Michael   7
Hewison, R.   107
Hick, Frank   48, 51, 52, 56–7
hierarchy   29
  managerial   46
  organisational, sophisticated   49
  straightforward   165
Hobbes, Thomas   25
Hobsbawm, Eric   21, 130
Hoggart, Richard   90
Hollowell, P.   23, 39, 88, 91, 98, 99
Hoole, Bill   99
Hornsea   81
hospitals   174
Hotels   44
Hotline   131
hours of work   26, 39
  appalling   38
  Government-imposed regulation
    18
  overtime   93
  unsociable   93
Howe, Bob   76, 77
Howell, D.   19, 25–6, 28, 45, 46, 70,
    92
Hudson, Mark   90
Hull Botanic Gardens shed   73
Hull Trains   165
Hunts Bank Offices   95
hydraulic and pumping-engine staff
    74

ideal types   39
ideology of work   10
IEA (Institute of Economic Affairs)
    124
IMF (International Monetary Fund)
    5
income   93
  see also wages
individuality   39, 85
  crushed   104
industrial relations   36, 114, 137
  breaking down of collective
    structures   124
  corporatist   113
  highly formalised   79
Industrial Revolution   99–100
industrialisation   82

inefficiencies 5, 44
inflation 5
infrastructure 16, 54, 149, 151, 169, 174
  ownership of 148
insecurity 11, 80, 146, 148, 151, 166
  chronic 154
  new levels of 171
  product of 156
  temporal 171
inspectors 23, 31, 32
  signalling 95, 96
instability 26, 100, 109
  temporal 155
Interbrand 127
Intercity 119
  East Coast 127, 128
  West Coast 120
interest groups 4
interest payments 48
intransigence 36
investment 54
  ability to fund 169
  borrowing capital for 48
  industry in need of 174
  new 19
  redundant 56
  speculative 16
Irving, R.J. 17, 23

Jack, Ian 139, 149
Japan 7
Jarvis Fast Line 165
jobs for life 25, 41
Johnson, Bill 63
joint stock companies 15

Kay, John 169
Kellett, J.R. 13, 90
Kennedy, A. 106, 152
Kings Cross 99, 127
Kingsford, P. W. 21, 22, 99

labour 9, 14, 25, 70, 99
  aristocracy of 167
  casual 149
  cutting turnover 24
  'dignified' 101
  direct clash between capital and 50
  direct control of 78
  forward march of 38
  general recognition of 39
  highly intensive 74
  immigrant 88, 98
  inability to recruit 86
  manual, representation of 101
  moral control over 35
  outsourcing of 148
  recruited and released 26–7
  role played by 85
  semi-skilled 151
  skilled 151
  strategic power of 72
  subcontracting 149
  super-aristocracy of 21
  timeless golden age of 87
  unrest 23
  unskilled 92–3
  see also division of labour;
    employment
labour markets 135
  formation of skill in 76
  internal 49
  newly fragmented 148
  tight 72, 78, 86, 93, 98, 148
labour policies 22, 24
  authoritarian 35
Labour Government 2, 4, 5, 6, 45, 47, 70
  commitment to public ownership 44
  new priorities for transport policy 64
Labour Party 3
  Blair's interpretation of past failure of 174
  class background of leading intellectuals in 47
  economic strategy 45
  elitist fear of 'workers' within 47
Ladbroke Grove 167, 169, 170
laissez-faire 16, 173
Leeds 155, 164
Le Goff, Jacques 128
Leicester 56
Leviathan 26, 35–9

liberal politics 36
liberalisation 124, 146, 155
liveries 41, 128, 130
  and logos 127
Liverpool 15
Liverpool Street Station 104
Livewire 127, 128, 130
LMR (London Midland Region) 71
LMS (London Midland and Scottish)
    17–18, 29, 44, 46, 70, 71
LNER (London North Eastern
    Railway) 18, 44, 49, 56, 65, 104,
    113, 127, 128
LNWR (London North Western
    Railway Company) 15
Loadhaul freight company 12, 124,
    125–6, 127, 137, 140, 141
Lockwood, D. 91
Locomotive Journal 32
London 14, 56, 96, 147, 164
  Victorian 25
  wartime 53
  see also Charing Cross; Euston;
        Kings Cross; Ladbroke Grove;
        Liverpool Street; LMR; LMS;
        LNER: LNWR; LPTB; Nine Elms;
        Paddington; Southall
London and Birmingham Railway
    19
London, Brighton and South Coast
    Railway 21
London Transport 44, 93, 96
London Transport Museum 167
London Underground 12, 93, 116,
    123, 142, 155, 157, 174
  staff made redundant 139
  training 153
Lowenthal, D. 108
loyalty 15, 24, 89, 125, 156, 158
  outraged 82
LPTB (London Passenger Transport
    Board) 4, 70
Lynn-Meek, V. 8, 106, 152

McDonalds 142
MacGregor, John 121–2, 171–2
McKenna, F. 19, 21, 22, 25, 28, 29,
    88–9, 99
McKibbin, Ross 48

McKillop, Norman 37–9, 69, 70, 72,
    112
McKinsey 64
Macmillan, Harold 5, 118, 119
maintenance 74
  lack of 19
  subcontracted 165
  track 148, 149, 151
Major, John 117–18, 119, 168, 171,
    172
management 47–54
  arbitrary power restricted 39
  autocratic 24
  challenged 35
  change in attitude of 141
  confrontational stance towards 37
  efficient 7
  embeddedness of 52, 62, 176
  forced to share power with
      supervisors 9
  high degree of control 15
  interference 1
  lampoons of 36
  macho 151
  new type of 84
  older and newer 109–10
  poor 7
  private monopoly 43
  relationship between workers and
      4
  resistance to 143
  sector 115
  seen as agents of change 8
  strong military flavour found in
      41
  tightening of control 25
  use of dependency to control
      workforce 25
  use of nostalgia 110
managerialism 7, 8
Manchester 15, 56, 95
manufacturing 16
marginalisation 9, 98, 124
  deliberate 146
market forces 5, 6, 11, 122, 172
  greater penetration of 147
  introduction of 148, 152, 156
marketing 55
Marples, Ernest 57–8, 59

Marris, P.   102, 138, 161
Mars bars   93–4
Marx, Karl   14, 19–21, 100, 167
Marxists   135
Mason, Frank   30–1
mastery   34
Mawhinney, Brian   122
Mead, Margaret   137, 159–60, 161
Meakin, David   100
Mengel, Philip   164
mergers   17
Metal Box   160
Metropolitan Line   96
Metzgar, Jack   159
MICs (Mutual Improvement Classes) 31–2
Midland Railway   25–6
militant syndicalism   47
military language   22
Ministry of Transport   56
mismanagement   170
Missenden, Eustace   46–7
modernisation   60, 82, 85, 88, 91, 93
   and deskilling   74–80
   enterprising plans for   71
   upheaval of   67
modernity   14, 15, 100, 127
monopolies   6, 43, 114, 146
   de facto   17
   entrenched   121
   possibility of   16
moral identity   11
moral order   140
morale   82, 118
   loss of   151
Morris, William   167
Morris Motors   92, 93
Morrison, Herbert   4, 43, 70
Morse single needle telegraph   51
motor industry *see* car industry
Mowbray, Morris   77–8, 79, 97
Mulligan, Walter   30, 81
Mumby, D.   61
municipal tramways   18
Munro, R.   108–9, 110, 114, 123, 125
Murphy, B.   112
Muthesius, S.   60
mutual improvement   29

National Association of Railway Workers   86
National Coal Board   3
national interest   4
nationalisation   2, 7, 19, 42, 43–66, 104
   commitment to   3
   financial arrangements of   48
   gauging early opinion about   85
   nostalgia for   164–77
   opposition to   4–5
   sympathy for   94
natural wastage   98
navvies   19
negligence   14
neocorporatism   5
neo-liberals   7, 8, 104
   'think tanks'   117
NER (North Eastern Railway)   23, 26, 36, 32, 48, 49
Network Rail   177
Network Southeast   119
new entrants   28, 72, 157
New Labour   104, 122–3, 173, 174
New Statesman and Nation, The   4
new technology   67, 78, 79, 80, 175
Newbould, D. A.   81
Newcastle   26, 67, 69, 93
Nichols, Theo   85, 99, 109, 175
Nicholson, Robert   23
Nine Elms   1, 96
norms   41, 96, 142
   established   135
   of reciprocity   61
North American railroad practice 23–4
North Eastern Region   64
Northern Spirit   126
Northumberland   72
nostalgia   1, 2, 8, 13, 85, 100, 105, 106, 116, 118, 119, 130, 164–77
   celebration of   131
   increased interest in   109
   interpretive   102
   management use of   110
   past times are impregnated with 107–8
   reflective   102
   romantic   81, 90

nostophobia 109, 110, 116, 126, 173
Nottingham 56
nuclear families 137
NUR (National Union of Railwaymen)
    46, 69

O for Q (Organising for Quality)
    scheme 115
O'Connell Davidson, J. 160
Old Oak Common 93
Oliver, John 139–40
Orwell, George 6, 118
overtime 93, 147
Oxford 118

Paddington 61, 88, 93, 118
Paget, Cecil 26
Panorama (BBC television) 120, 121
Parker, M. 8, 134
parliamentary select committees
    16–17, 26
part-time workers 26
passenger journeys 18
paternalism 24, 41
    autocratic 35
Patten, Chris 117
pay differentials 92
Payton, P. 81
Pearson, A.J. 46, 60
Pelaw 31
Pelling, H. 45
penalty payments 165
Pendleton, A. 46, 79, 112
Penn, R. 29, 73, 76
Pennsylvania 24
pensions 24, 93
permanence 85, 100
Perry, G. 85
Peters, T. 105, 152, 160
platelayers 23
Pollins, H. 18, 44
Pollitt, C. 7
Porter, John 140, 143
porters 21, 87
post-figurative culture 137, 160
Powell, D. 70
power 154
    arbitrary exercise of 25
    devolving to regional level 54

shared 9
technologies of 10
unequal 135
predictability 100, 122, 133, 138,
    161
price regulation 18
pride 125, 151
private sector 3, 6, 16
    aims to liberate from State
        restrictions 111
    faith in the power of 6
privatisation 5, 6, 44, 54, 105,
    116–23
    biggest surprise of 164
    effects of 2
    labelled a disaster 170
    loss of staff 153
    one of the major ironies of 146
    one of the political ideas behind
        156
    trying to make sense of 11
    uncertainty created by 151
problem-solving 77
production 10, 21, 136
    mass 101
Production-Engineering 64
productivity 5, 45, 66, 78
    closure proclaimed on the grounds
        of 85
    poor 7
    search for greater levels of 112
profitability 18
profits 26, 56
promotion 24, 25, 41, 72
    based on seniority 22, 116
    opportunities for 92
    rapid 23, 95, 96, 99
    seriously curtailed prospects 154
    static 88
Provincial 119
PSBR (Public Sector Borrowing
    Requirement) 5, 111, 168–9
public accountability 4, 123
public good 44
public opinion 16, 36
public ownership 3, 48
    commitment to 4, 44, 45
    early disillusion with 92
    hostility to 46

public sector
  crisis and Government failure   6
  denigration of   2
  failures worked in   110
  managerialism in   7
  solutions for the problems of   166
  transition to   48
  see also PSBR
public utilities   6
Putnam, L.   61

radical liberals   3
Rail (magazine)   124, 127, 164
Rail Regulator   165
Railtrack   119, 122–3, 131, 140, 144, 165, 168, 177
  ability to fund investment   169
  company-wide improvement programme   126
  competitive bidding for contracts   149
  highly overvalued   169
  incompetence of senior figures in   167
  move to restore public confidence   166
  operational responsibilities   176
  ownership of track and infrastructure   148
  refusal to extend Government funding for   170
  speed restrictions   164
  top 100 people changed   167
  see also Edmonds
Railway Act (1844)   18, 45
railway architecture   60
Railway Clerks' Association   39
Railway Executive   44, 45–7, 48, 54
Railway Magazine, The   126
Railways Act (1993)   116
Ramsay, H.   143, 160
Ranger, T.   130
rationalisation   2, 17, 21, 44, 56, 61, 67, 74, 88, 93
  community effects of   81
  impact of   80
Rawlinson, Robert   19
Raymond, Sir Stanley   61, 63–4
Raynes, J.R.   37, 112

'Rebecca myth'   110
recession   18, 92
recruitment   15, 21, 27, 49, 67, 93, 124, 166
  bureaucratic   22
  difficulties   94
  from the street   142, 149
  from universities   84
  non-white New Commonwealth labour   88
  wartime   98
red tape   2, 104
redundancy   26, 27, 59, 126, 139, 146, 147, 154, 166, 170, 171
  threat of   25
  voluntary   125, 141, 142
Rees, R.   3, 6
reforms   44, 82, 114
refresher courses   3
Reid, Bob   113–14, 120
renationalisation   169, 170
reorganisation   19
repairs   74, 169
  massive cost of   19
Report on the Reshaping of British Railways (1963)   58
reputation   30, 31
respect   51, 97, 140, 141
responsibility   93
  career   156
  clear lines of   165
  moral   96
restrictive practices   7
restructuring   11, 18, 92, 113, 116, 147, 152, 153
  encouraging   2
  industrial   9
  inevitable   66
  lost opportunities for   6
  marginalisation as a result of   91
  massive, industry in need of   174
retirement   23, 99, 140
  early   139, 142, 154, 171
retrenchments   25
revelry   19
Revill, G.   22, 24, 99
reward   2, 24
Reynolds, Michael   28
Ridley, Nicholas   117

Rifkin, J.   10
risks   149
RMT (National Union of Rail, Maritime and Transport Workers)   142, 149, 151, 154–5
road transport   18, 44, 55, 111
Roberts, I.   90, 125, 160
Robinson, Sir Brian   54
Rolt, L.T.C.   55–6, 82
romance   81, 85, 90, 120, 128, 130
   classic   131
ROSCOs (rolling stock leasing companies)   165
rosters   79, 89, 112, 113, 124–5
Rothbury   72
routes   14, 16
   rival   17
Rowbotham, Sheila   173
Rugby   56
rule books   15, 22, 25
Ruskin, John   167

safety   17, 36, 51, 149, 151
Salaman, G.   89, 90, 91, 97, 98, 99
salaries   15, 147
Samuel, Raphael   7, 35, 60, 104, 108, 118, 173
Savage, M.   15, 22, 23, 24, 39
Saville, R.   3, 4, 5
schools   174
Scot Rail   120
Scotland   34, 156, 164
Sea Containers   127
Seabrook, Jeremy   90
Seacontainers   156
seasonality   25, 26
second men   89, 98
sectorisation   113–16
security   24, 41, 82, 89, 92, 155–6
   assured   19
   freedom surrendered in return for   25
   limited   171
   long-term   93, 152
   ontological   109, 110
Seldon, A.   117
self-help   29, 77
self-preservation   25

Semmens. P.W.B.   99
seniority   23, 96, 140
   compensation by   24
   downplaying the importance of   84
   promotion based on   22, 116
   reduction in   95
Sennett, Richard   11, 138, 160, 161, 175
sentimentality   112, 117, 119
service ethic   2
severance   147
   voluntary   125, 157
Shaw, J.   122
shed culture   30
shedmen   74
Sheffield   56
Shinwell, Emmanuel (Manny)   3
shipbuilding/shipbuilders   18, 101
shipyard 'fathers'   34
short-termism   175
shunters   74
signalling   16, 51, 74, 79, 166
   semaphore   82
   technical change in   79
signalmen   33–4, 36, 71, 74, 79, 80, 81, 92, 93, 140, 143
   pride in being   157
   relief   96
   trainee   95
Simmons, Jack   36
Sisson, K.   116
skill levels   21, 79, 153–4
   disparity in   148
   implications for   91
Slim, General Sir William   45
Slough   93
Smith, M.   41, 42
Smith, P.W.   81
social change   102
social housing programmes   90
social order   29, 161
social relations   82, 159, 176
socialisation   54, 73, 81
   early or anticipatory   50
   traditional forms of   137
   training and   28–35, 41
socialists   3, 101
   young   71

solidarity 10, 158
Southall 167, 169, 170, 171, 176
Southern Region 96
SR (Southern Railway) 17, 44, 46,
    47, 51
stability 26, 34, 35, 54, 67, 81, 96,
    99
  byword for 19
  challenge to 66
  loss of 11, 100
  threat posed to 25
Stagecoach 156
standardisation 78–9
State ownership
  active opposition to 47
  attack on 87
  interpreted 70
  met with general acclaim 69
  three themes of 85
station masters 118, 141
status 23, 29, 67, 72, 79, 85, 90, 141
  changing 89
  loss of 69, 88, 91–102
  non-skilled 41
  skilled 37
steam preservation 90
steamraisers 74
Stedman Jones, G. 25
steel 5, 18, 101, 159
Stein, M.B. 16, 154
stereotypes 39, 41, 87, 101
Stokes, Sir John 118, 119
storekeepers 74
Storey, John 115, 116, 160
Storey, Tom 31, 32
Stourbridge 69
Strangleman, T. 12, 26, 90, 121,
  126, 160
strikes 70–1, 112
subcontractors 165
Summerseat 84
Sunderland 26
superintendents 26
supervisors 15, 48–9, 50, 59, 97,
  140
  change in attitude of 141
  loss of respect from 96
Swann, Paul 101
Swindon 16, 118

SWT (South West Trains) 147
Sykes, R. 81

Talyllyn line 55–6
Taylor, A.J. 16
TCLs (train crew leaders) 140
team-building 2
technical/technological change 19,
  79, 80, 98
Teesside 80
Tennant (NER General Manager) 26
Tenniel, John 36
tension 7, 14, 21, 59, 63, 138, 152,
  172
  urban and rural 101
tenure 24
Terkel, S. 97
Thatcher, Margaret 2, 6–7, 69,
  110–11, 117, 118, 172
  first administration 43
  privatisation 5, 6
Thatcherism 5, 7
'theatres of memory' 35
threats 25, 26
timekeepers 74
timelessness 100
Titfield Thunderbolt, The 55, 85–6
TOCs (train operating companies)
  148, 155, 165
Tolstoy, L.N. 100
Tomlinson, J. 4, 70, 78
Tönnies, Ferdinand 100
TOUs (Train Operating Units) 119,
  120
track 16, 148, 149, 151
trade unions 5, 32, 41–2, 44, 69,
  100, 112
  breaking the power of 124
  footplate 79
  impact of 15
  managers cowed by overpowerful
    dinosaurs 104
  official policy 70
  see also ASLEF; ASRS; General
    Union; National Association of
    Railway Workers; NUR; Railway
    Clerks' Association; RMT; TSSA;
    United Pointsmen's and
    Signalmen's Society

traditions 128, 131
  ingrained 7
  strong 15, 28
Train, John Landale 45–6
Train Circular 53
Train Crew Agreement 124
training 1, 3, 15, 54
  costs of 154
  management 49
  need to formalise 153
  non-driving railway employees 148
  socialisation and 28–35, 41
Trainload Freight 127
Transport Acts (1962/1968) 57, 64
Travers, T. 98, 117, 124, 155
TSSA (Transport Salaried Staffs' Association) 121, 144, 146
Turner, Arthur 29, 100
Turner, J.M.W. 14
turnover of staff 93, 94, 95, 98
  discouraged 24
Tweedmouth 23
Tyne 26, 76, 77

Uff, John 171, 176
underinvestment 134
unemployment 5, 93
unfilled vacancies 93
United Pointsmen's and Signalmen's Society 39
unsafe practices 149
upskilling 76
Urwick, Orr and Partners 64
US railroad/steel industry 97, 159
utopia 105

values 42, 87, 134, 135
  family 7
  shared 41
  younger, fresher 130
Variable Rostering 124–5
Vaughan, A. 81–4, 95
vertical integration 16, 24, 121
viaducts 17
Victorian era 7, 15, 21, 23, 25, 118
  power of industrial Leviathans 26
  prospects for working-class men 41

railway management challenged 35
  unions 36
Virgin 126, 130–1, 148, 151, 164, 167

wages 17, 21–2, 71, 86, 94
  ability to move to follow 156
  disparity in 92
  downward pressure on 66
  Government-imposed regulation 18
  indifferent 25
  low 24
  renewed resistance to increases 92
Wales 15, 55–6
Warburton, Ivor 120
water 5, 6, 122, 160
water-softening plant attendants 74
Waterman, R.H. 105, 152
Webb, Sidney and Beatrice 47
Weber, Max 21, 100
Wedderburn, D. 64
welfare 24, 39, 43
welfare cheats 7
Welfare State 173
Welsby, John 117, 123, 124
West Anglia Great Northern 165
West Auckland 31
West Indians 88
Western Region 71, 78, 81, 82, 118
white-collar workers 49, 73, 171
  divisions between blue-collar and 144
Williams, K. 6
Williams, R. 41, 95, 100, 101
Willmott, P. 91, 152
Wilson, Harold 48, 60, 64
'winter of discontent' (1978–9) 6, 111, 174
Winterton, J. 46
Withernsea 81
Wolmar, C. 117, 123, 131, 172
work ethic 2, 167
work practices 2, 9, 124
  modern, anachronistic barrier to 112
working class 19, 41, 73, 89, 90, 124
  king of 97

working conditions   25, 38, 141,
      142
   deterioration of   146, 171
   decline in   92
   disparity in   92
World War II   19, 34, 51, 52, 72, 73,
      95, 98
   rail chaos autumn and winter
         (2000) compared to   166

Wright, Patrick   107
Wright, S.   5, 8, 105
Wright Mills, C.   59

York   51, 52
Young, Gilly   72–3
Young, M.   91

Zweig, Ferdynand   134